DIMENSIONAL ANALYSIS
FOR UNIT CONVERSION
USING MATLAB®

DIMENSIONAL ANALYSIS FOR UNIT CONVERSION USING *MATLAB*®

Roger W. Pryor, PhD

MERCURY LEARNING AND INFORMATION
Dulles, Virginia
Boston, Massachusetts
New Delhi

Publisher: David Pallai
MERCURY LEARNING AND INFORMATION
22841 Quicksilver Drive
Dulles, VA 20166
info@merclearning.com
www.merclearning.com
(800) 232-0223

Roger W. Pryor. *Dimensional Analysis for Unit Conversion Using MATLAB®*.
ISBN: 978-1-683922-42-1

Printed on acid-free paper in the United States of America.

The publisher recognizes and respects all marks used by companies, manufacturers, and developers as a means to distinguish their products. All brand names and product names mentioned in this book are trademarks or service marks of their respective companies. Any omission or misuse (of any kind) of service marks or trademarks, etc. is not an attempt to infringe on the property of others.

Library of Congress Control Number: 2018943744
181920321

Our titles are available for adoption, license, or bulk purchase by associations, universities, corporations, etc. Digital versions of this title may be purchased at www.authorcloudware.com or other e-vendors. *DRM protected disc files are available for download from the publisher by writing to info@merclearning. com.* For additional information, please contact the Customer Service Dept. at 1-(800) 232-0223 (toll free) or info@merclearning.com.

The sole obligation of MERCURY LEARNING AND INFORMATION to the purchaser is to replace the disc, based on defective materials or faulty workmanship, but not based on the operation or functionality of the product.

CONTENTS

PREFACE

This app converts values from one unit of measure to another using standard conversion factors. It performs conversions from and to the inch-pound system units used in the United States of America and the International System of Units (SI) as documented in the National Institute of Standards and Technology (NIST) publications of conversions for general use. [1,2] There are 1,316 conversion factors available for bidirectional conversion from/to SI units, organized into 44 minor subsections by topic under eight major topical sections. There is also an alphabetical section comprising 445 conversion factors for unidirectional conversion to SI units.

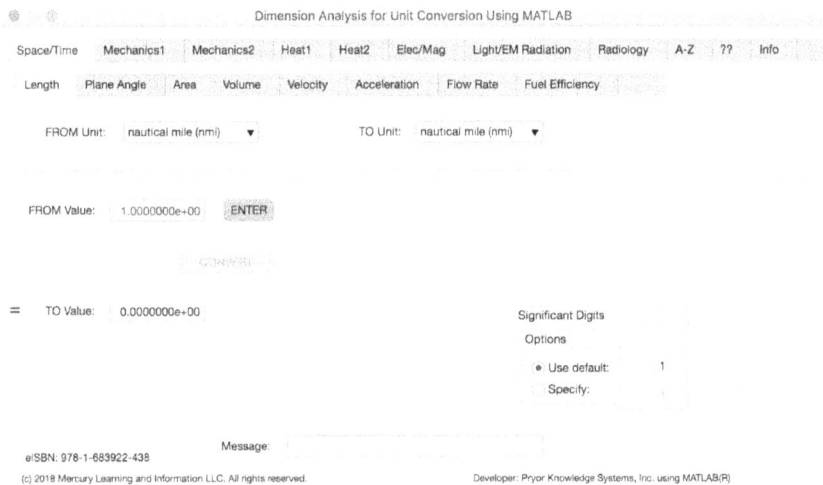

			Dimension Analysis for Unit Conversion Using MATLAB									
Space/Time	Mechanics1	Mechanics2	Heat1	Heat2	Elec/Mag	Light/EM Radiation		Radiology	A-Z	??	Info	
Length	Plane Angle	Area	Volume	Velocity	Acceleration	Flow Rate	Fuel Efficiency					

FROM Unit: nautical mile (nmi) ▾   TO Unit: nautical mile (nmi) ▾

FROM Value: 1.0000000e+00   ENTER

CONVERT

=   TO Value: 0.0000000e+00

Significant Digits

Options
- Use default: 1
- Specify:

Message:

eISBN: 978-1-683922-438

Developer: Pryor Knowledge Systems, Inc. using MATLAB(R)

The application performs all three steps in the conversion process: application of the relevant conversion factor, selection of significant digits, and rounding of the result. Conversion factors designated as "exact" are definitions, or they have been set by agreements that define the factor value precisely. All other conversion factors, designated as "derived," result from truncation of decimal places and/or calculation by a combination of other factors.

INPUT VARIABLES

- Units of measure

    a. Select the major category of units to be converted using the tab group at the top of the application window.

    b. Select the minor category of units to be converted using the lower tab group.

    c. Use the Left pull-down list to select units to be converted (From) and use the Right pull-down list to select units to be converted (To).

- See Appendix B for a complete list of available categories.

ABOUT THE DEVELOPERS

Pryor Knowledge Systems, Inc. offers the highest quality of technological services: consulting, multiphysics model design and training, and application development. We enjoy providing superior service to our customers, and we have assisted them in achieving their goals since 1993. Contact: *rwpryor@pksez1.com*. Website: *www.pksez1.com*.

The President and CEO, **Roger W. Pryor, Ph.D**., a scientist with an extensive background, has an international reputation in materials research, electronics technology and electronics applications in both industry and academia. Dr. Pryor ran his own technical consulting firm, R. W. Pryor and Associates, before merging with Pryor Knowledge Systems in 2003. He was a Professor, Research, at Wayne State University for 14 years, leading research teams in the development and characterization of semiconductor materials and devices. Dr. Pryor also has an industrial background; having led advanced research teams at Energy Conversion Devices in Troy, Mich., and at Pitney Bowes in Norwalk, Conn., and served as a member of the technical staff at Bell Laboratories in Whippany, New Jersey. He received his M.S. and Ph. D. degrees in Physics from the Pennsylvania State University after earning a B.S. in Physics from Worcester Polytechnic Institute. Dr. Pryor holds 24 patents, primarily in semiconductor devices. He has written four college textbooks, the most recent of which is *Multiphysics Modeling Using COMSOL 5 and MATLAB*, published in 2015. He served in the United States Navy.

The founder of Pryor Knowledge Systems in 1993, **Beverly E. Pryor**, Vice President and COO, has developed application systems and databases supporting industrial research, administrative, and manufacturing functions for platforms ranging from mainframes to microcomputers and hand-held devices. Ms. Pryor managed IT departments at Trinova Corp., Vickers, Inc., Pitney Bowes Credit Corporation, and Allied Corporation, and programmed systems at Pennsylvania State University and Paul Revere Insurance Company.

CHAPTER 1

INTRODUCTION TO DIMENSIONS

1.1   WHAT ARE DIMENSIONS?

A first order approximation of the definition of a dimension is that it is a measurable extent, such as length, width, depth, and so forth {1}. The mental concept that initiated the first attempt at understanding the creation of a basic set of dimensions initially arose after one of our early wandering ancestors had acquired both an adequate amount of spare energy and sufficient inquisitive desire to attempt trying to comprehend and record the puzzle of dimensions as well as the ongoing demands of surviving each day in their hostile life-threatening environment.

This successful ancestor's leisure became available after the ancestor's group developed sufficient foresight and ability to obtain, preserve, and defend surplus quantities of foodstuffs. This required the group to have the ability to defeat, fend-off, or out-think other hungry predator groups, both animal and human. The first true level of new understanding of the functioning of their immediate environment came about slowly when different individuals within the group developed special (unique) skills. Those skilled individuals in each group gradually developed improvements in their intuitive skills. The net result was that each skilled individual aided in raising the overall growth of the long-term survival rate for all the members of their group.

As each skilled individual survived and then, through cultural transmission, passed on their newly acquired observational and survival skills to the next generation, they also passed on the desire to survive to each of the subsequent generations. Those members and groups that did not participate in the skill acquisition and transfer process failed to survive and are not now here. We, the readers of this book, are among the present surviving members of this skill-based process.

One of the first dimensional concepts that impinged on the leisure time consciousness of our earlier ancestors was the concept of time {2}. At least one of some group's more observant individuals noted that the local environment often became lighter (day {3}) and darker (night {4}). At a later point, some of those same individuals and perhaps their friends additionally noted that during the darker period (night) various patterns comprising lighted {5} objects (moon {6}, stars {7}, planets {8}, meteors {9}, comets {10}, etc.) were visible only during the dark period. Some of said individuals also noted that a few of the patterns were relatively constant, whereas other patterns moved rapidly and some patterns moved very slowly across the dark (sky). The observers additionally noted that the object that they called the moon had a repeating pattern (currently named a lunar month {11}(approximately 30 days)), of light and dark cycles.

Once one of these early ancestors had an initial grasp of the concept of time, the next dimensional concept that this curious ancestor would have also noted, perhaps simultaneously, is that of the change of the environmental temperature {12}. As each observer's time flowed onward, the light level in the local environment rose and fell with the transition through the day-night cycle. At some point, the early observer would also have noted that the local environment would, over a period of moons (several days = month) get warmer and later get cooler. As the number of lunar months passed, the temperature of the local environment would fluctuate from hotter to colder cyclically (summer {13}, winter {14}) during the yearly {15} seasons {16}. The observation of environmental temperature fluctuations to lower values (cooling) would, in some locations, promote the discovery and use of fire {17}, the development of clothing {18}, and eventually the development of shelters {19}. Perhaps the group would have to migrate to a warmer environment if they knew what dimensions to use to guide travel to that warmer location.

All of the preceding observations require stored memories (either in the head or carved into wood, rock, clay, or metal) of earlier environments and their conditions. They also require that the recorder have sufficient caloric {20} intake and the ability to learn writing {21} and counting {22}. Sufficient caloric intake implies cooking {23} and language {24} skills, and those imply that the earlier ancestor have an adequate brain size {25} with sufficient structural complexity for rational {26} thinking {27}.

1.2 WHY IS AN UNDERSTANDING OF DIMENSIONS AND THEIR USAGE NECESSARY?

Even the most fundamental survival skills require a basic intuitive under-standing of the concept of and application of dimensional analysis to our everyday tasks. We need to consider such concepts as size (bigger > same > smaller), flow (faster > same > slower), etc. The universe that we live in requires both a large range and a diverse collection of different dimensional types (see Figure 1-1). In order to understand the workings of this universe, at least to First Order (minimally), we humans must be able to recognize and utilize the relative influence and relative importance of applied dimen-sional analysis. We must also consider the effects that modifications and/or errors in applied dimensional values will have on the events and opportuni-ties occurring in our real lives on a timely (second-by-second) basis.

Figure 1-1 {28} shows a quick overview of the role of applied dimensional analysis considerations that are required in characterizing the dimensions of various objects that are of interest in this universe. In this case, it is shown by employing the CGS (centimeter (cm), gram (g), second (s)) system to indicate relative size of various well known objects. In this book, only Classical Physics {29} examples are considered. For clarity, however, the location of the non-Classical Black Hole Region and non-Classical Quantum Region are also shown in Figure 1-1.

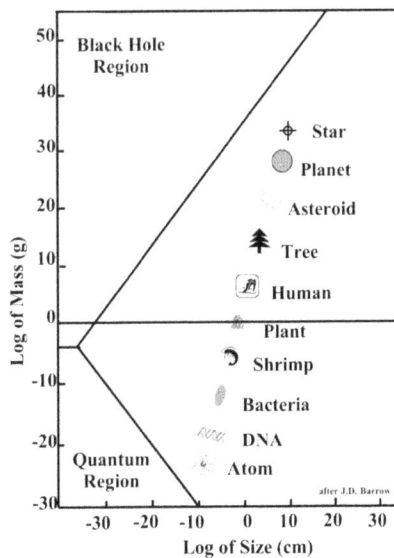

FIGURE 1-1 Constant Atomic Density Size Plot.

A BASIC INTRODUCTION TO DIMENSIONAL ANALYSIS APPLIED TO UNIT CONVERSION

2.1   WHAT ARE THE BASE AND THE DERIVED UNITS PRESENTED HEREIN?

A physical quantity comprises a numerical value (e.g., 43) and an associated unit (i.e., meters). The aforementioned physical quantity constitutes one of the base SI units (meter) with the amplitude as defined by the attached numerical value (43). In the SI Unit System there are seven (7) base units. The SI Base Units are as shown in Table 2-1 {30}:

TABLE 2-1

Quantity	Unit Name	Symbol
length	meter	m
mass[a]	kilogram	kg
time	second	s
electric current	ampere	A
thermodynamic temperature	kelvin	K
amount of substance	mole	mol
luminous intensity	candela	cd
(a) United States standard practice for laws and regulations, "weight" is a synonym for mass.		

The SI derived units used in science, engineering, and mathematics disciplines are formed by combining the base units algebraically. The use of derived units is to improve the ease of calculation and to allow for improved problem concept understanding. In the case of commonly used derived units, the International System of Units has assigned special names to 22 of the most frequently used derived units, two of which are the dimensionless units (radian (rad) and steradian (sr)) (see Table 2-2 {31}).

TABLE 2-2

Function	Special Name	Special Symbol	Expression In Non-Base SI Units	Expression In Base SI Units
plane angle	radian	rad	1	$m * m^{-1}$
solid angle	steradian	sr	1	$m^2 * m^{-2}$
frequency	hertz	Hz	-	s^{-1}
force	newton	N	-	$m * kg * s^{-2}$
pressure	pascal	Pa	$N * m^{-2}$	$m^{-1} * kg * s^{-2}$
energy	joule	J	$N * m$	$m^2 * kg * s^{-2}$
power	watt	W	$J * s^{-1}$	$m^2 * kg * s^{-3}$
electric charge	coulomb	C	-	$s * A$
electromotive force	volt	V	$W * A^{-1}$	$m^2 * kg * s^{-3} * A^{-1}$
capacitance	farad	F	$C * V^{-1}$	$m^{-2} * kg^{-1} * s^4 * A^2$
resistance	ohm	Ω	$V * A^{-1}$	$m^2 * kg * s^{-3} * A^{-2}$
conductance	siemens	S	$A * V^{-1}$	$m^{-2} * kg^{-1} * s^3 * A^2$
magnetic flux	weber	Wb	$V * s$	$m^2 * kg * s^{-2} * A^{-1}$
magnetic flux density	tesla	T	$Wb * m^{-2}$	$kg * s^{-2} * A^{-1}$
inductance	henry	H	$Wb * A^{-1}$	$m^2 * kg * s^{-2} * A^{-2}$
Celsius temperature	degree Celsius[a]	°C	-	K
luminous flux	lumen	lm	$cd * sr$	cd
illuminance	lux	lx	$lm * m^{-2}$	$cd * m^{-2}$
radionuclide activity	becquerel	Bq	-	s^{-1}
absorbed dose, kerma[b]	gray	Gy	$J * kg^{-1}$	$m^2 * s^{-2}$

(Continued)

Function	Special Name	Special Symbol	Expression In Non-Base SI Units	Expression In Base SI Units
dose equivalent	sievert	Sv	$J \circ kg^{-1}$	$m^2 \circ s^{-2}$
catalytic activity	katal	kat	-	$mol \circ s^{-1}$

(a) The degree Celsius is the special name for the kelvin used to express Celsius temperatures. The degree Celsius and the kelvin are equal in size, so that the numerical value of a temperature difference or temperature interval is the same when expressed by either set of units.

(b) kinetic energy released per unit mass

The rules of physical analysis and algebra require that for all dimensional mathematical equations {32}, the results of the combined resolved {33} elements on each side of the equality statement (=) must be numerically and unitarily {34} coherent {35} (see Example 1).

Linear Velocity Example 1:

$$1\frac{in}{s} = 1\frac{\cancel{in}}{s} * 2.54\frac{cm}{\cancel{in}} = 2.54\frac{cm}{s} \tag{1}$$

where:

The linear dimension of 1 inch in the Inch-Pound Units system is defined as being equal to 2.54 centimeters in the SI system and the magnitude of the unit second (s) is equal in both systems. In the process of conversion, the conversion factor cancels the redundant unit factors (in and 1/in), as shown.

CHAPTER 3

AN INTRODUCTION TO THE INTERNATIONAL SYSTEM OF UNITS (SI) – Conversion Factors for General Use (Bidirectional Conversion App Section)

In the NIST Special Publication 1038{36}, the International System of Units (SI) – Conversion Factors for General Use are defined. Conversion factors are the mathematical multiplier or divisor and associated units that allow any compatible measurements to be converted from one system of units to another system of units (e.g., inches to centimeters or vice versa). As with any measurement, caution needs to be applied so that resolution (number of digits) is not confused with accuracy (number of significant digits). The general rule of thumb (first approximation) is to use no more digits than are absolutely necessary. See the Appendix for more details on significant digits.

The examples below demonstrate calculations as applied to obtain the results shown by the Unit Convertor MATLAB App provided with this book.

3.1 QUANTITIES OF SPACE/TIME

Figure 3.1-0 shows the Unit Converter MATLAB App Front Panel with the Space/Time Tab selected. Bidirectional matrices were created and installed within this app to allow the user the ability to easily select via pull-down menus all of the desired conversion pairs required. Each of the Tabs available in the App is explored herein.

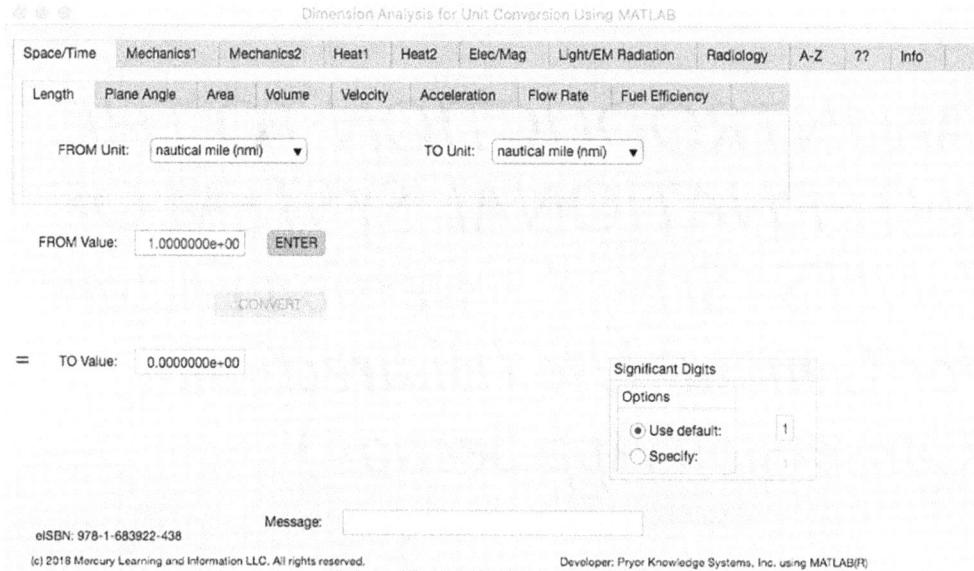

FIGURE 3.1-0 Unit Converter Using MATLAB App - Front Panel - Space/Time Tab.

Length Tab

The first Minor Tab under the Space/Time Major Tab is the Length Tab (see Figure 3.1-1). This example demonstrates the conversion of Nautical Miles to Fathoms. In this example, the amplitude of Nautical Miles is set equal to one (1.0). The conversion equation is shown in Example 3.1-1 {37}.

Nautical Miles (nmi) (=1.0) to Fathoms (ftm) Example 3.1-1:

ftm = nmi * km/nmi * m/km * ftm/m = nmi * 1.852 * 1000 * 1/1.828804
= nmi * 1.01268370E3

Where: ftm = the number of fathoms calculated
nmi = the number of nautical miles to be converted (1.0)
km/nmi = the conversion factor from nautical miles to kilometers (1.852)
m/km = the conversion factor from kilometers to meters (1000)
ftm/m = the conversion factor from meters to fathoms (1/1.828804)

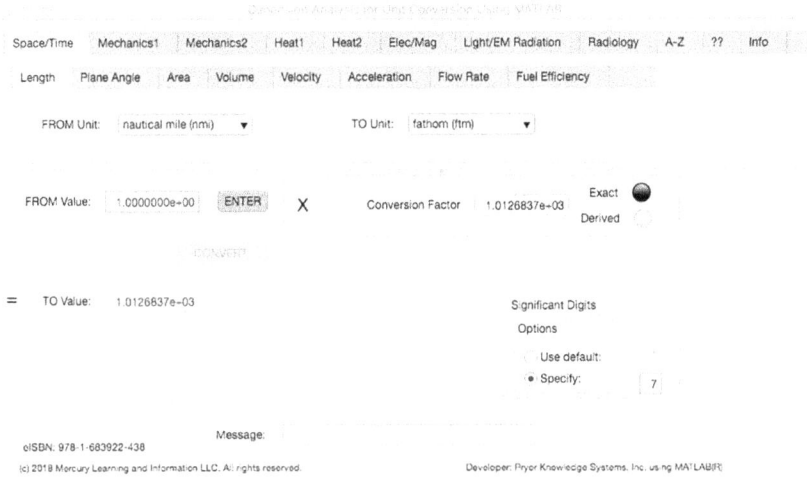

FIGURE 3.1-1 Unit Converter Using MATLAB App - - Space/Time Length Tab.

And:

The resulting solution is:

ftm = 1.0 * 1.01268370E3 = 1.01268370E3 Fathoms   (3.1-1)

The Space/Time Length Tab allows the bidirectional selection of the following length dimensions:

Major-Tab	Minor-Tab	Units on Pull-Down
Space/Time		
	Length	
		nautical mile (nmi)
		mile (mi)
		kilometer (km)
		fathom (ftm)
		yard (yd)
		meter (m)
		foot (ft)
		foot (U.S. Survey)
		inch (in)
		centimeter (cm)
		millimeter (mm)
		pica, printer's (12p)
		point, printer's (p)

Major-Tab	**Minor-Tab**	**Units on Pull-Down**
Space/Time		
	Length	
		mil (0.001 in)
		micrometer (µm)
		microinch (µin)
		nanometer (nm)
		angstrom

Plane Angle Tab

The second Minor Tab under the Space/Time Major Tab is the Plane Angle Tab (see Figure 3.1-2). This example shows the conversion of Radian to degree arc. In this example, the amplitude of Radian is set equal to one (1.0). The conversion equation is shown in Example 3.1-2.

Radian (rad) to degree arc (°) Example 3.1-2:

degree arc = rad ° degrees/rad = rad ° 57.29578 = 57.29578 degrees

Where: degree arc = the number of degrees calculated
rad = the starting number of Radian (1.0)
degrees/rad = the conversion factor from Radian to degrees (57.29578)

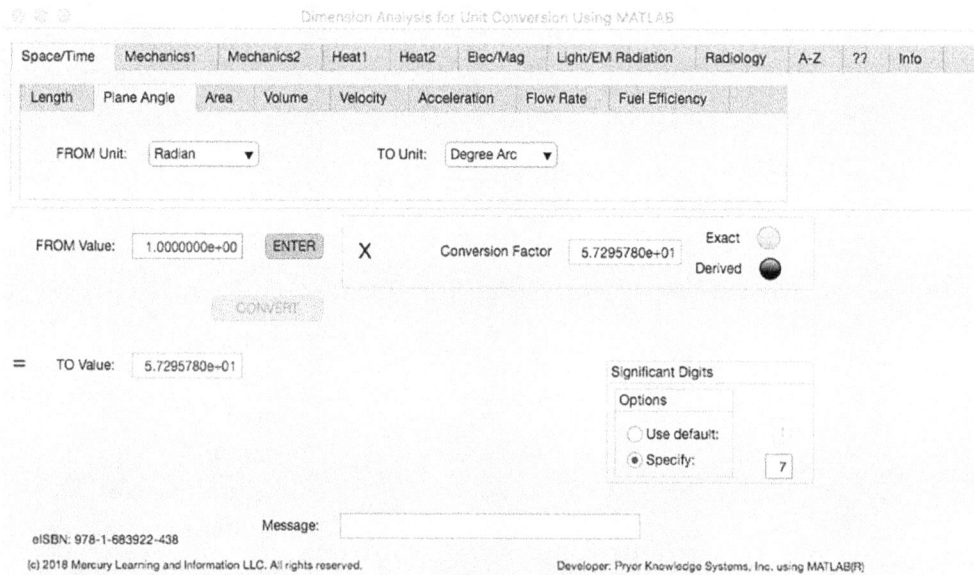

FIGURE 3.1-2 Unit Converter Using MATLAB App - Space/Time Plane Angle Tab.

And:

The resulting solution is:

$$\text{degree arc} = 1.0 \text{ ° } 57.29578 = 57.29578 \text{ degrees} \qquad (3.1\text{-}2)$$

The Space/Time Plane Angle Tab allows the bidirectional selection of the following angle dimensions:

Major-Tab	**Minor-Tab**	**Units on Pull-Down**
Space/Time		
	Plane Angle	
		Radian
		Degree Arc

Area Tab

The third Minor Tab under the Space/Time Major Tab is the Area Tab (see Figure 3.1-3). This example shows the conversion of Square Feet to Square Kilometers. In this example, the amplitude of Square Feet is set equal to one (1.0). The conversion equation is shown in Example 3.1-3.

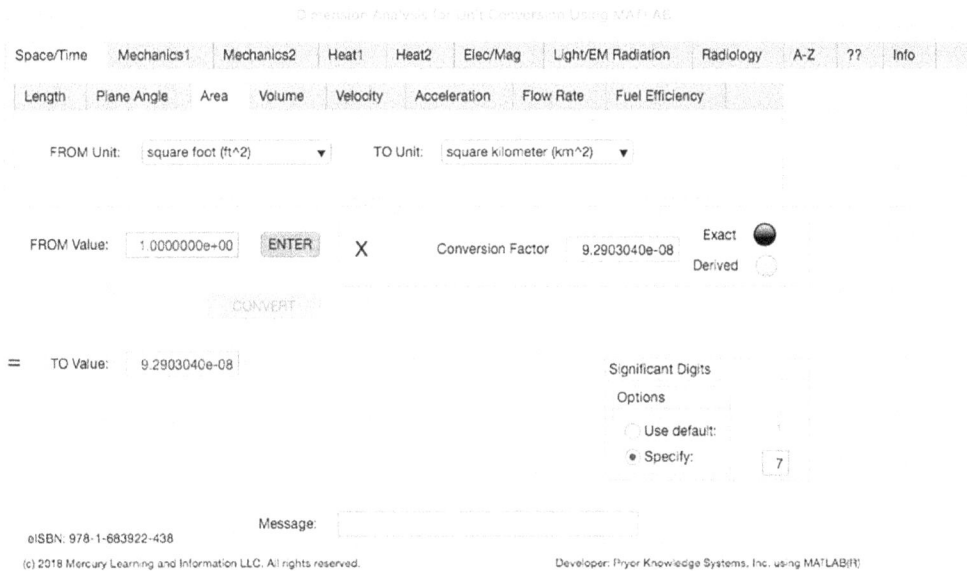

FIGURE 3.1-3 Unit Converter Using MATLAB App - Space/Time Area Tab.

Square Feet (ft^2) (=1.0) to Square Kilometers (km^2) Example 3.1-3:

$$km^2 = ft^2 * m^2/ ft^2 * km^2/ m^2 = ft^2 * 9.290304E\text{-}2 * 1/ 1.0E6$$
$$= ft^2 * 9.290304E\text{-}8$$

Where: km^2 = the number of Square Kilometers calculated

ft^2 = the number of Square Feet to be converted (1.0)

m^2/ft^2 = the conversion factor from Square Feet to Square Meters (9.290304E-2)

km^2/m^2 = the conversion factor from Square Meters to Square Kilometers (1/ 1.0E6)

And:

The resulting solution is:

$$km^2 = 1.0 * 9.290304E\text{-}8 = 9.290304E\text{-}8 \text{ Square Kilometers} \qquad (3.1\text{-}3)$$

The Space/Time Area Tab allows the bidirectional selection of the following area dimensions:

Major-Tab	Minor-Tab	Units on Pull-Down
Space/Time		
	Area	
		square mi (mi^2)
		square km (km^2)
		hectare (ha)
		acre
		square m (m^2)
		square yd (yd^2)
		square ft (ft^2)
		square in (in^2)
		square cm (cm^2)
		square mm (mm^2)
		circular mil

Volume Tab

The fourth Minor Tab under the Space/Time Major Tab is the Volume Tab (see Figure 3.1-4). This example shows the conversion of Acre-Foot to Liter. In this example, the amplitude of Acre-Foot is set equal to one (1.0). The conversion equation is shown in Example 3.1-4.

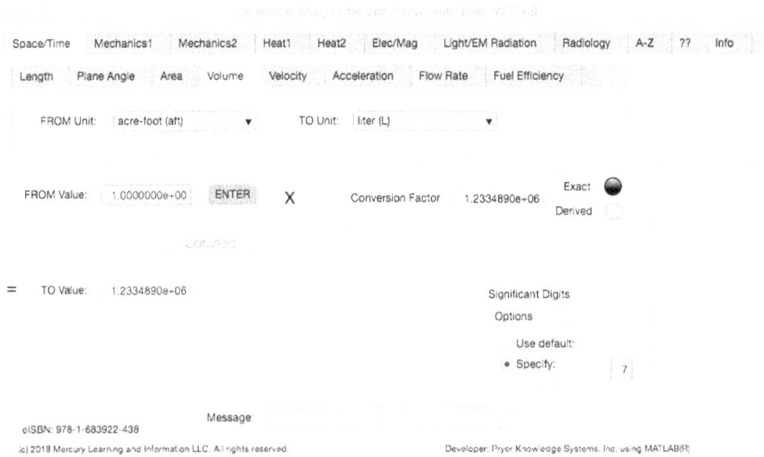

FIGURE 3.1-4 Unit Converter Using MATLAB App - Space/Time Volume Tab.

Acre-Foot (aft) (=1.0) to Liter (L) Example 3.1-4:

$$L = aft * m^3/aft * L/m^3 = aft * 1.233489E3 * 1 / 1.0E-3 \{38\}$$
$$= aft * 1.23348900E6$$

Where: L = the number of Liter calculated
      aft = the number of Acre-Foot to be converted (1.0)
    m^3/aft^3 = the conversion factor from Acre-Foot to Cubic Meters
        (1.233489E3)
     L/m^3 = the conversion factor from Cubic Meters to Liter (1/ 1.0E-3)

And:

The resulting solution is:

$$L = 1.0 * 1.23348900E6 = 1.23348900E6 \text{ Liter} \tag{3.1-4}$$

The Space/Time Volume Tab allows the bidirectional selection of the following volume dimensions:

Major-Tab	Minor-Tab	Units on Pull-Down
Space/Time		
	Volume	
		acre-foot (aft)
		barrel-oil (bbl)
		cubic-yard (yd^3)
		cubic-meter (m^3)

Major-Tab	**Minor-Tab**	**Units on Pull-Down**
Space/Time		
	Volume	
		cubic-foot (ft^3)
		board-foot (BF)
		register-ton (RT)
		bushel (bu)
		gallon (gal)
		liter (L)
		quart (qt)
		pint (pt)
		fluid-ounce (fl oz)
		milliliter (mL)
		cubic-inch (in^3)
		cubic-centimeter (cm^3)

Velocity Tab

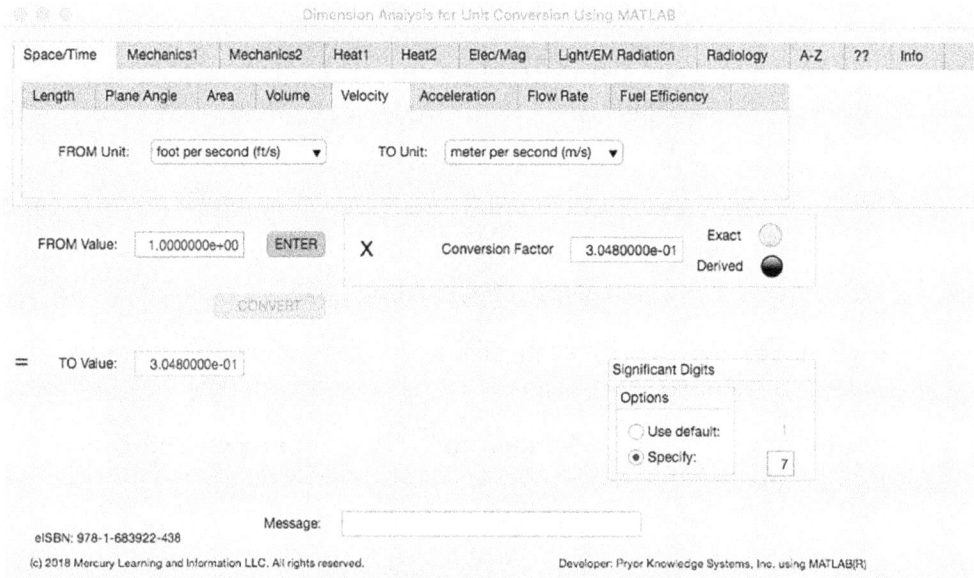

FIGURE 3.1-5 Unit Converter Using MATLAB App - Space/Time Velocity Tab.

The fifth Minor Tab under the Space/Time Major Tab is the Velocity Tab (see Figure 3.1-5). This example shows the conversion of Foot per Second to Meter per Second. In this example, the amplitude of Foot per Second (ft/s) is set equal to one (1.0). The conversion equation is shown in Example 3.1-5.

Foot per Second (ft/s) (=1.0) to Meter per Second (m/s) Example 3.1-5:

> m/s = ft/s * m/ft = ft/s * 0.3048

Where: m/s = the number of Meter per Second calculated
       ft/s = the number of Foot per Second to be converted (1.0)
       m/ft = the conversion factor from Foot to Meter (0.3048)

And:

> The resulting solution is:

> m/s = 1.0 * 0.3048 = 0.3048 Meter per Second   (3.1-5)

The Space/Time Velocity Tab allows the bidirectional selection of the following velocity dimensions:

Major-Tab	Minor-Tab	Units on Pull-Down
Space/Time		
	Velocity	
		foot per second (ft/s)
		mile per hour (mi/hr)
		knot (nmi/hr)
		meter per second (m/s)
		kilometer per hour (km/hr)

Acceleration Tab

The sixth Minor Tab under the Space/Time Major Tab is the Acceleration Tab (see Figure 3.1-6). This example shows the conversion of Foot per Second Squared to Meter per Second Squared. In this example, the amplitude of Foot per Second Squared (ft/s^2) is set equal to one (1.0). The conversion equation is shown in Example 3.1-6.

FIGURE 3.1-6 Unit Converter Using MATLAB App - Space/Time Acceleration Tab.

Foot per Second Squared (ft/s²) (=1.0) to Meter per Second Squared (m/s²) Example 3.1-6:

$$m/s^2 = ft/s^2 * m/ft = ft/s^2 * 0.3048$$

Where: m/s² = the number of Meter per Second Squared calculated
ft/s² = the number of Foot per Second Squared to be converted (1.0)
m/ft = the conversion factor from Foot to Meter (0.3048)

And:

The resulting solution is:

$$m/s^2 = 1.0 * 0.3048 = 0.3048 \text{ Meter per Second Squared} \tag{3.1-6}$$

The Space/Time Acceleration Tab allows the bidirectional selection of the following acceleration dimensions:

Major-Tab	Minor-Tab	Units on Pull-Down
Space/Time		
	Acceleration	
		inch per second squared (in/s^2)
		foot per second squared (ft/s^2)
		meter per second squared (m/s^2)
		standard acceleration of gravity (g)

Flow Rate Tab

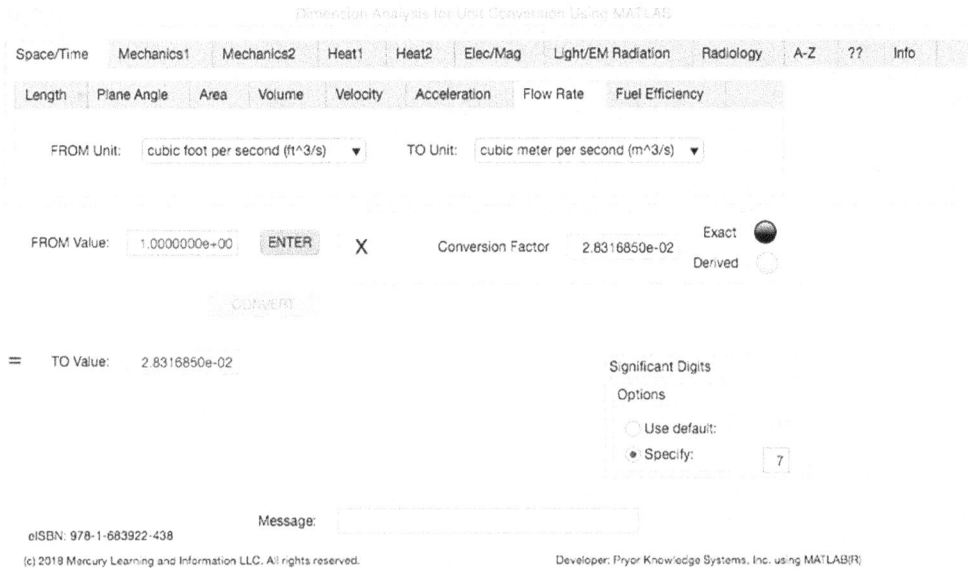

FIGURE 3.1-7 Unit Converter Using MATLAB App - Space/Time Flow Rate Tab.

The seventh Minor Tab under the Space/Time Major Tab is the Flow Rate Tab (see Figure 3.1-7). This example shows the conversion of Cubic Foot per Second to Cubic Meter per Second. In this example, the amplitude of Cubic Foot per Second (ft³/s) is set equal to one (1.0). The conversion equation is shown in Example 3.1-7.

Cubic Foot per Second (ft³/s) (=1.0) to Cubic Meter per Second (m³/s) Example 3.1-7:

$$m^3/s = ft^3/s * m^3/ft^3 = ft^3/s * 2.831685E\text{-}2$$

Where: m^3/s = the number of Cubic Meter per Second calculated
  ft^3/s = the number of Cubic Foot per Second to be converted (1.0)
  m^3/ft^3 = the conversion factor from Cubic Foot to Cubic Meter
  (2.831685E-2)

And:

The resulting solution is:

$m^3/s = 1.0 * 2.831685E\text{-}2 = 2.831685E\text{-}2$ cubic meter per second   (3.1-7)

The Space/Time Flow Rate Tab allows the bidirectional selection of the following flow rate dimensions:

Major-Tab	Minor-Tab	Units on Pull-Down
Space/Time		
	Flow Rate	
		cubic foot per second (ft^3/s)
		cubic foot per minute (ft^3/min)
		cubic yard per minute (yd^3/min)
		gallon per minute (gal/min)
		gallon per day (gal/day)
		cubic meter per second (m^3/s)
		liter per second (L/s)
		liter per day (L/day)

Fuel Efficiency Tab

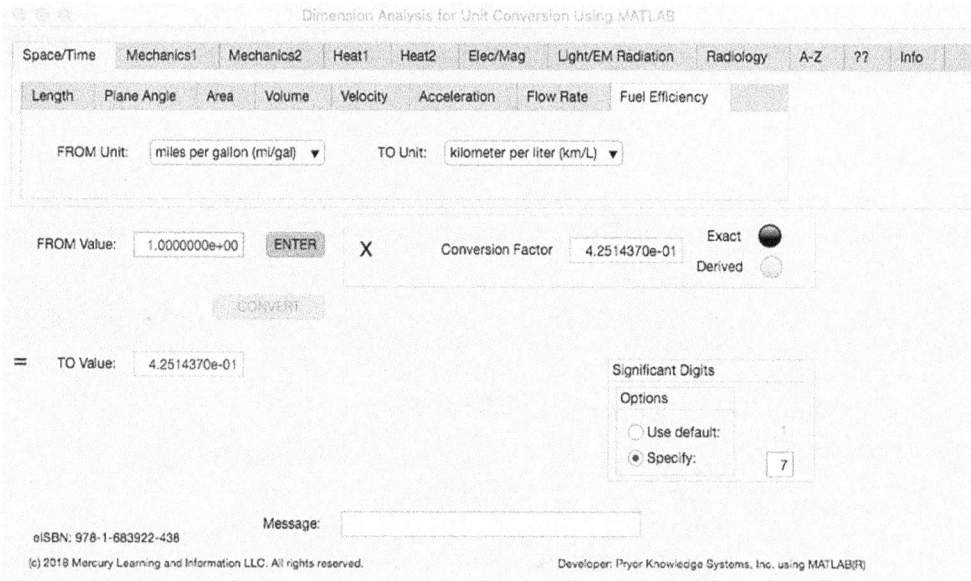

FIGURE 3.1-8 Unit Converter Using MATLAB App - Space/Time Fuel Efficiency Tab.

The eighth Minor Tab under the Space/Time Major Tab is the Fuel Efficiency Tab (see Figure 3.1-8). This example shows the conversion of Mile per Gallon to Kilometer per Liter. In this example, the amplitude of Mile per Gallon (mi/gal) is set equal to one (1.0). The conversion equation is shown in Example 3.1-8.

Mile per Gallon (mi/gal) (=1.0) to Kilometer per Liter (km/L) Example 3.1-8:

$$km/L = mi/gal * km/mi * gal/L = mi/gal * 1.609344 * 1/3.785412$$
$$= mi/gal * 4.251437E\text{-}1$$

Where: km/L = the number of Kilometer per Liter calculated
    mi/gal = the number of Mile per Gallon to be converted (1.0)
    km/mi = the conversion factor from Miles to Kilometers (1.609344)
    gal/L = the conversion factor from Liter to Gallon (1/3.785412)

And:

The resulting solution is:

$$km/L = 1.0 * 4.251437E\text{-}1 = 4.251437E\text{-}1 \text{ Kilometer per Liter} \quad (3.1\text{-}8)$$

The Space/Time Fuel Efficiency Tab allows the bidirectional selection of the following fuel efficiency dimensions:

Major-Tab	Minor-Tab	Units on Pull-Down
Space/Time		
	Fuel Efficiency	
		miles per gallon (mi/gal)
		kilometer per liter (km/L)

3.2 QUANTITIES OF MECHANICS1

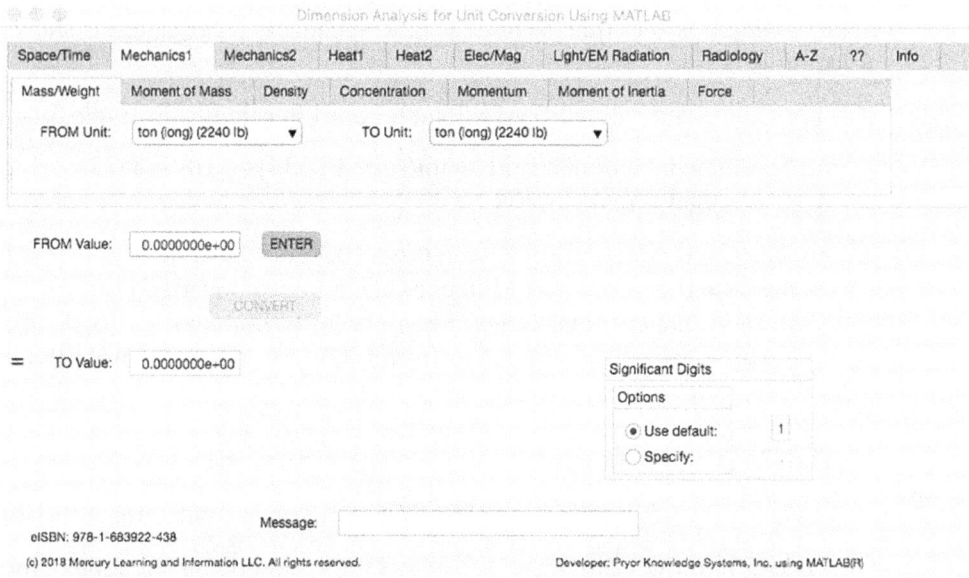

FIGURE 3.2-0 Unit Converter Using MATLAB App - Front Panel Mechanics1 Tab.

Figure 3.2-0 shows the Unit Converter MATLAB App Front Panel with the Mechanics1 Major Tab selected. Bidirectional matrices were created and installed within this app to allow the user the ability to easily select via pull-down menus all of the desired conversion pairs required. Each of the Mechanics1 Minor Tabs available in the App is explored herein.

Mass/Weight Tab

The first Minor Tab under the Mechanics1 Major Tab is the Mass/Weight Tab (see Figure 3.2-1). This example demonstrates the conversion of ton (long) (2240 lb) to ounce (avoirdupois). In this example, the amplitude of ton (long) is set equal to one (1.0). The conversion equation is shown in Example 3.2-1.

ton (long) (2240 lb) (=1.0) to ounce (avoirdupois) (oz) Example 3.2-1:

$$oz = ton(long) * lb/ton(long) * oz/lb = 1.0 * 2240 * 16$$
$$= ton (long) * 3.584E4$$

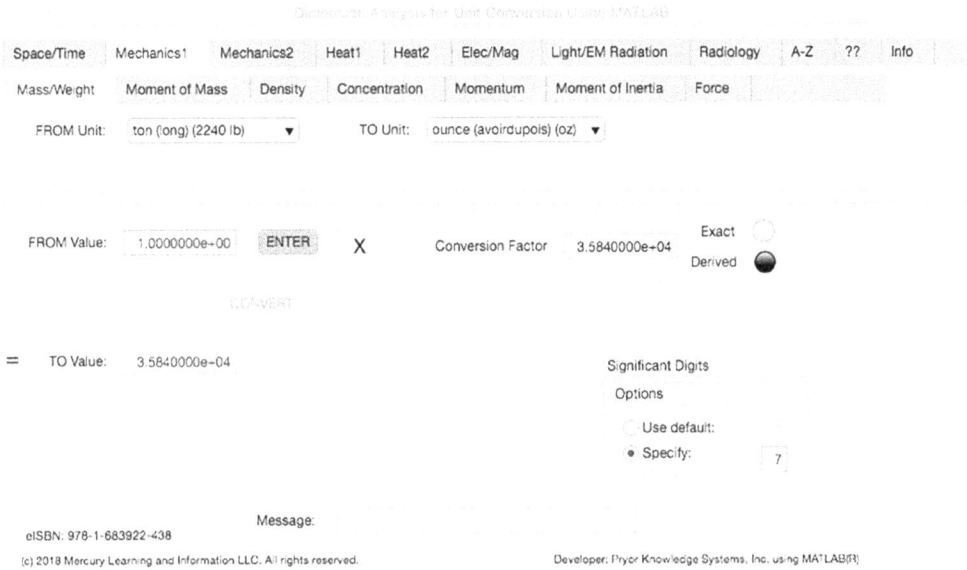

FIGURE 3.2-1 Unit Converter Using MATLAB App - Mechanics1 Mass/Weight Tab.

Where: oz = the number of ounces calculated
ton (long) = the number of tons to be converted (1.0)
lb/ton (long) = the conversion factor from ton (long) to pounds (2240)
oz/lb = the conversion factor from pounds to ounces (16)

And:

The resulting solution is:

$$oz = ton\ (long) * 3.584E4 = 3.584E4\ ounces \qquad (3.2\text{-}1)$$

The Mechanics1 Mass/Weight Tab allows the bidirectional selection of the following mass/weight dimensions:

Major-Tab	Minor-Tab	Units on Pull-Down
Mechanics1		
	Mass/Weight	
		ton (long) (2240 lb)
		metric ton (t)
		ton (short) (2000 lb)

Major-Tab	Minor-Tab	Units on Pull-Down
Mechanics1		
	Mass/Weight	
		slug
		kilogram (kg)
		pound (avoirdupois)
		ounce (troy)
		ounce (avoirdupois)
		gram (g)
		grain
		milligram (mg)

Moment of Mass Tab

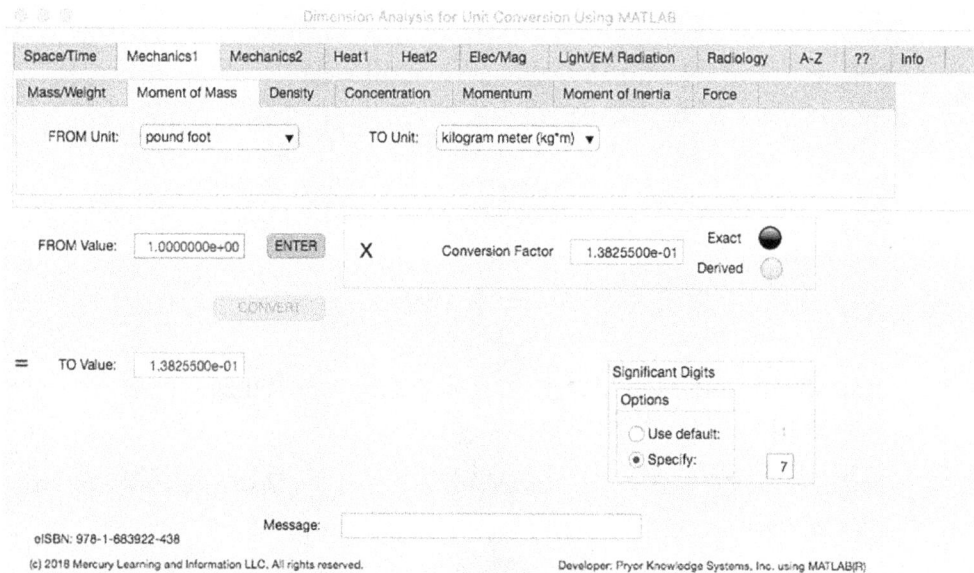

FIGURE 3.2-2 Unit Converter Using MATLAB App - Mechanics1 Moment of Mass Tab.

The second Minor Tab under the Mechanics1 Major Tab is the Moment of Mass Tab (see Figure 3.2-2). This example demonstrates the conversion of pound foot to kilogram meter. In this example, the amplitude of pound foot (lb°ft) is set equal to one (1.0). The conversion equation is shown in Example 3.2-2.

pound foot (lb°ft) (=1.0) to kilogram meter (kg°m) Example 3.2-2:

$$kg°m = lb°ft * kg/lb * m/ft = 1.0 * 0.45359252 * 0.3048$$
$$= lb°ft * 1.38255E\text{-}1$$

Where: kg°m = the number of kilogram meter calculated
    lb°ft = the number of pound foot to be converted (1.0)
    kg/lb = the conversion factor from pound to kilogram (0.45359252)
    m/ft = the conversion factor from foot to meter (0.3048)

And:

The resulting solution is:

$$kg°m = 1.0 * 1.38255E\text{-}1 = 1.38255E\text{-}1 \text{ kilogram meter} \qquad (3.2\text{-}2)$$

The Mechanics1 Moment of Mass Tab allows the bidirectional selection of the following Moment of Mass dimensions:

Major-Tab	Minor-Tab	Units on Pull-Down
Mechanics1		
	Moment of Mass	
		pound foot
		kilogram meter (kg ° m)

Density Tab

The third Minor Tab under the Mechanics1 Major Tab is the Density Tab (see Figure 3.2-3). This example demonstrates the conversion of pound per cubic foot to kilogram per cubic meter. In this example, the amplitude of pound per cubic foot (lb/ft^3) is set equal to one (1.0). The conversion equation is shown in Example 3.2-3.

pound per cubic foot (lb/ft^3) (=1.0) to kilogram per cubic meter (kg/m^3) Example 3.2-3:

$$kg/m^3 = lb/ft^3 * kg/lb * ft^3/m^3 = 1.0 * 0.45359252 * 3.5314662E1$$
$$= lb/ft^3 * 1.601846E1$$

Where: kg/m^3 = the number of kilogram per cubic meter calculated
lb/ft^3 = the number of pound per cubic foot to be
converted (1.0)
kg/lb = the conversion factor from pound to kilogram
(0.45359252)
ft^3/m^3 = the conversion factor from cubic meter to cubic foot
(3.5314662E1)

And:

The resulting solution is:

kg/m^3 = lb/ft^3 * 1.601846E1= 1.601846E1 kilogram
per cubic meter

(3.2-3)

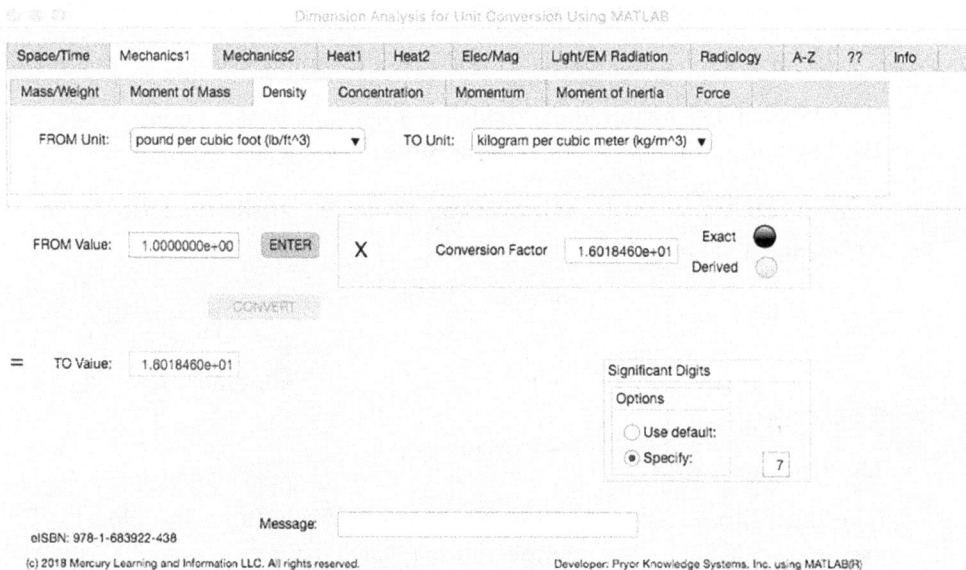

FIGURE 3.2-3 Unit Converter Using MATLAB App - Mechanics1 Density Tab.

The Mechanics1 Density Tab allows the bidirectional selection of the following Density dimensions:

Major-Tab	**Minor-Tab**	**Units on Pull-Down**
Mechanics1		
	Density	

ton (2000 lb [short]) per cubic yard

metric ton per cubic meter (t/m^2)

pound per cubic foot (lb/ft^3)

kilogram per cubic meter (kg/m^3)

Concentration Tab

The fourth Minor Tab under the Mechanics1 Major Tab is the Concentration Tab (see Figure 3.2-4). This example demonstrates the conversion of pound per gallon to gram per liter. In this example, the amplitude of pound per gallon (lb/gal) is set equal to one (1.0). The conversion equation is shown in Example 3.2-4.

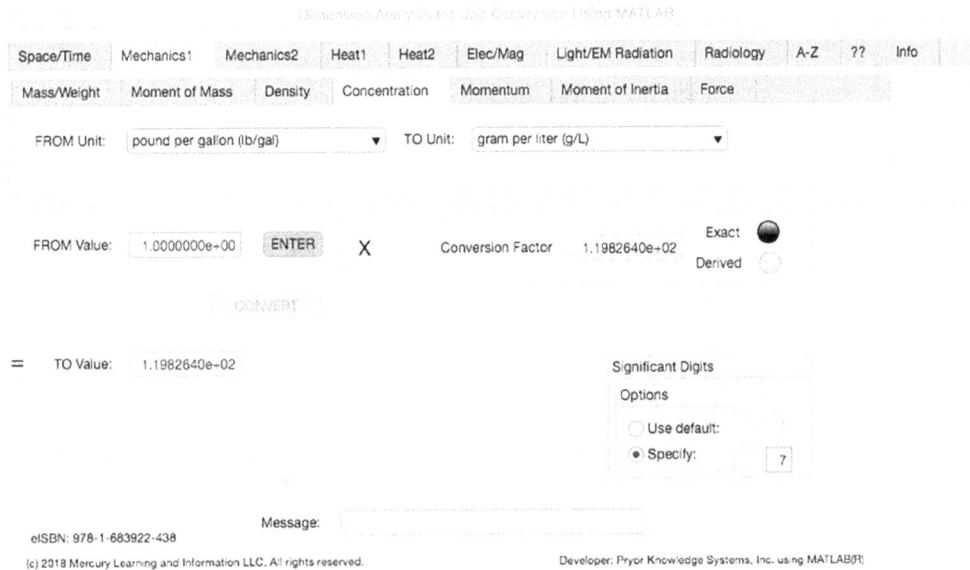

FIGURE 3.2-4 Unit Converter Using MATLAB App - Mechanics1 Concentration Tab.

pound per gallon (lb/gal) (=1.0) to gram per liter (g/L) Example 3.2-4:

g/L = lb/gal * kg/lb * g/kg * gal/L
    = 1.0 * 0.45359237 * 1.0E3 * 1/ 3.785412
    = lb/gal * 1.1982642E2

Where: g/L = the number of gram per liter calculated
       lb/gal = the number of pound per gallon to be converted (1.0)
       kg/lb = the conversion factor from pound to kilogram (0.45359252)
       g/kg = the conversion factor from kilogram to gram (1.0E3)
       gal/L = the conversion factor from liter to gallon (1/ 3.785412)

And:

The resulting solution is:

g/L = lb/gal * 1.1982642E2= 1.1982642E2 gram per liter   (3.2-4)

The Mechanics1 Concentration Tab allows the bidirectional selection of the following Concentration dimensions:

Major-Tab	Minor-Tab	Units on Pull-Down
Mechanics1		
	Concentration (mass)	
		pound per gallon (lb/gal)
		ounce (avoirdupois) per gallon (oz/gal)
		gram per liter (g/L)

Momentum Tab

The fifth Minor Tab under the Mechanics1 Major Tab is the Momentum Tab (see Figure 3.2-5). This example demonstrates the conversion of pound foot per second to kilogram meter per second. In this example, the amplitude of pound foot per second (lb*ft/s) is set equal to one (1.0). The conversion equation is shown in Example 3.2-5.

pound foot per second (lb*ft/s) (=1.0) to kilogram meter per second (kg*m/s) Example 3.2-5:

kg*m/s = lb*ft/s * kg/lb * m/ft = lb*ft/s * 4.5359252E-1 * 3.048E-1
       = lb*ft/s * 1.38255E-1

Where: kg°m/s = the number of kilogram meter per second calculated
lb°ft/s = the number of pound foot per second to be converted (1.0)
kg/lb = the conversion factor from pound to kilogram
(4.5359252E-1)
m/ft = the conversion factor from foot to meter (3.048E-1)

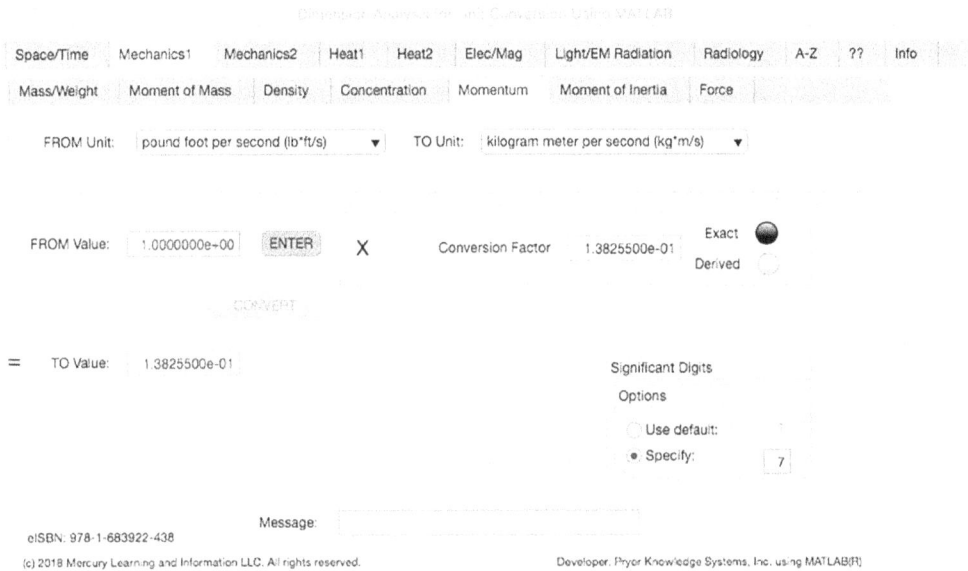

FIGURE 3.2-5 Unit Converter Using MATLAB App - Mechanics1 Momentum Tab.

And:

The resulting solution is:

kg°m/s = lb°ft/s ° 1.38255E-1 = 1.38255E-1 kilogram meter
per second   (3.2-5)

The Mechanics1 Momentum Tab allows the bidirectional selection of the following Momentum dimensions:

Major-Tab	Minor-Tab	Units on Pull-Down
Mechanics1		
	Momentum	
		pound foot per second (lb°ft/s)
		kilogram meter per second (kg°m/s)

Moment of Inertia Tab

The sixth Minor Tab under the Mechanics1 Major Tab is the Moment of Inertia Tab (see Figure 3.2-6). This example demonstrates the conversion of pound square foot to kilogram square meter. In this example, the amplitude of pound square foot (lb°ft^2) is set equal to one (1.0). The conversion equation is shown in Example 3.2-6.

pound square foot (lb°ft^2) (=1.0) to kilogram square meter (kg°m^2) Example 3.2-6:

$$kg°m^2 = lb°ft^2 * kg/lb * m^2/ft^2$$
$$= lb°ft^2 * 4.5359252E\text{-}1 * 9.290300E\text{-}2$$
$$= lb°ft^2 * 4.2140110E\text{-}2$$

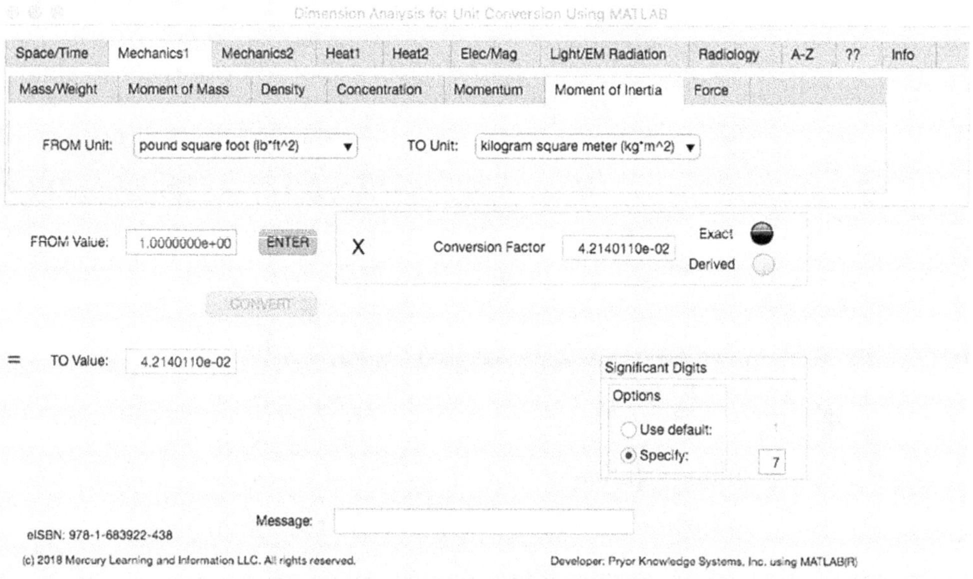

FIGURE 3.2-6 Unit Converter Using MATLAB App - Mechanics1 Moment of Inertia Tab.

Where: kg°m^2 = the number of kilogram meter squared calculated
      lb°ft^2 = the number of pound foot squared to be converted (1.0)
      kg/lb = the conversion factor from pound to kilogram
        (4.5359252E-1)

m^2/ft^2 = the conversion factor from foot squared to meter squared (9.290300E-2)

And:

The resulting solution is:

$kg \cdot m^2$ = 1.0 $\cdot$ 4.2140110E-2 = 4.2140110E-2 kilogram meter squared

(3.2-6)

The Mechanics1 Moment of Inertia Tab allows the bidirectional selection of the following Moment of Inertia dimensions:

Major-Tab	Minor-Tab	Units on Pull-Down
Mechanics1		
	Moment of Inertia	
		pound square foot ($lb \cdot ft^2$)
		kilogram square meter ($kg \cdot m^2$)

Force Tab

The seventh Minor Tab under the Mechanics1 Major Tab is the Force Tab (see Figure 3.2-7). This example demonstrates the conversion of pound-force to Newton (N). In this example, the amplitude of pound-force (lbf) is set equal to one (1.0). The conversion equation is shown in Example 3.2-7.

pound-force (lbf) (=1.0) to Newton (N) Example 3.2-7:

$$N = lbf \cdot kg/lb \cdot g_n$$
$$= lbf \cdot 4.5359237E\text{-}1 \cdot 9.80665$$
$$= lbf \cdot 4.4482216E0$$

Where: N = the number of Newton calculated
    lbf = the number of pound-force being converted (1.0)
    kg/lb = the conversion factor from pound to kilogram (4.5359237E-1)
    g_n = the gravitational acceleration at the surface of the earth (9.80665)

And:

The resulting solution is:

N = 1.0 * 4.4482216E0 Newton   (3.2-7)

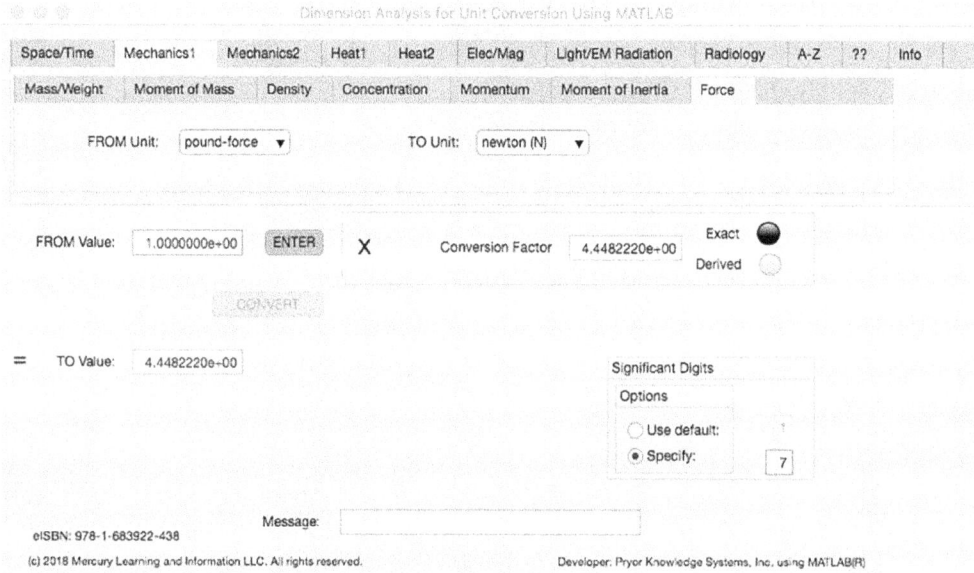

FIGURE 3.2-7 Unit Converter Using MATLAB App - Mechanics1 Force Tab.

The Mechanics1 Force Tab allows the bidirectional selection of the following Force dimensions:

Major-Tab	Minor-Tab	Units on Pull-Down
Mechanics1		
	Force	
		pound-force
		poundal
		newton (N)

3.3   QUANTITIES OF MECHANICS2

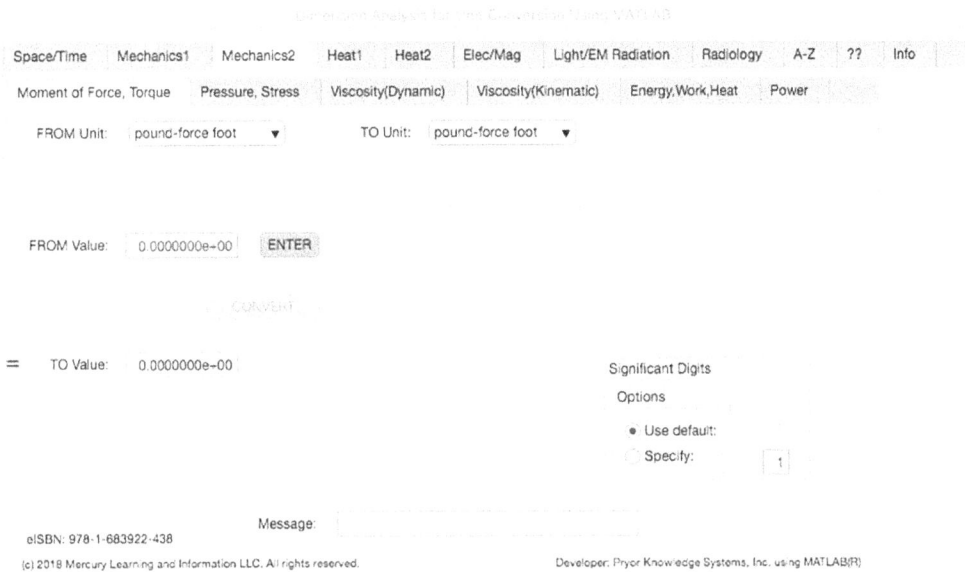

FIGURE 3.3-0 Unit Converter Using MATLAB App - Front Panel Mechanics2 Tab.

Figure 3.3-0 shows the Unit Converter MATLAB App Front Panel with the Mechanics2 Major Tab selected. Bidirectional matrices were created and installed within this app to allow the user the ability to easily select via pull-down menus all of the desired conversion pairs required. Each of the Mechanics2 Minor Tabs available in the App is explored herein.

Moment of Force, Torque Tab

The first Minor Tab under the Mechanics2 Major Tab is the Moment of Force, Torque Tab (see Figure 3.3-1). This example demonstrates the conversion of pound-force foot (lbf*ft) to Newton meter (N*m). In this example, the amplitude of pound-force foot (lbf*ft) is set equal to one (1.0). The conversion equation is shown in Example 3.3-1.

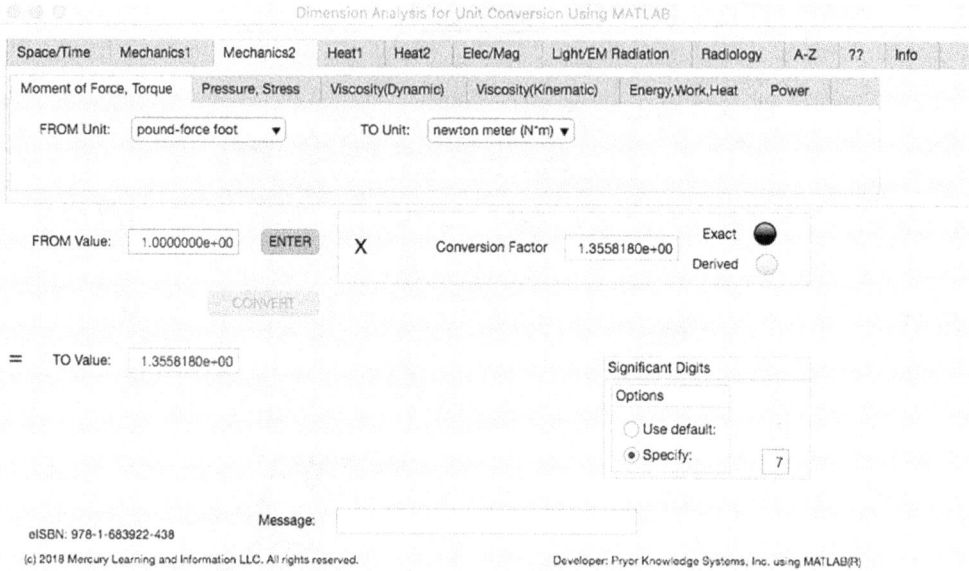

FIGURE 3.3-1 Unit Converter Using MATLAB App - Mechanics2 Moment of Force, Torque Tab.

pound-force (lbf) * foot (ft) (=1.0) to Newton (N) * meter (m) Example 3.3-1:

$$N*m = lbf * kg/lb * g_n *ft$$
$$= lbf * 4.5359237E\text{-}1 * 9.80665 * 0.3048$$
$$= lbf * ft * 1.355818E0$$

Where: N * m = the number of Newton meter calculated

lbf * ft = the number of pound-force foot being converted (1.0)

kg/lb = the conversion factor from pound to kilogram
(4.5359237E-1)

g_n = the gravitational acceleration at the surface of the earth
(9.80665)

m/ft = the conversion factor from foot to meter (0.3048)

And:

The resulting solution is:

$$N*m = 1.0 * 1.355818E0 = 1.355818E0 \text{ Newton-meter} \qquad (3.3\text{-}1)$$

The Mechanics2 Moment of Force, Torque Tab allows the bidirectional selection of the following Moment of Force, Torque dimensions:

Major-Tab	Minor-Tab	Units on Pull-Down
Mechanics2		
	Moment of Force, Torque	
		pound-force foot
		pound-force inch
		newton meter (N*m)

Pressure, Stress Tab

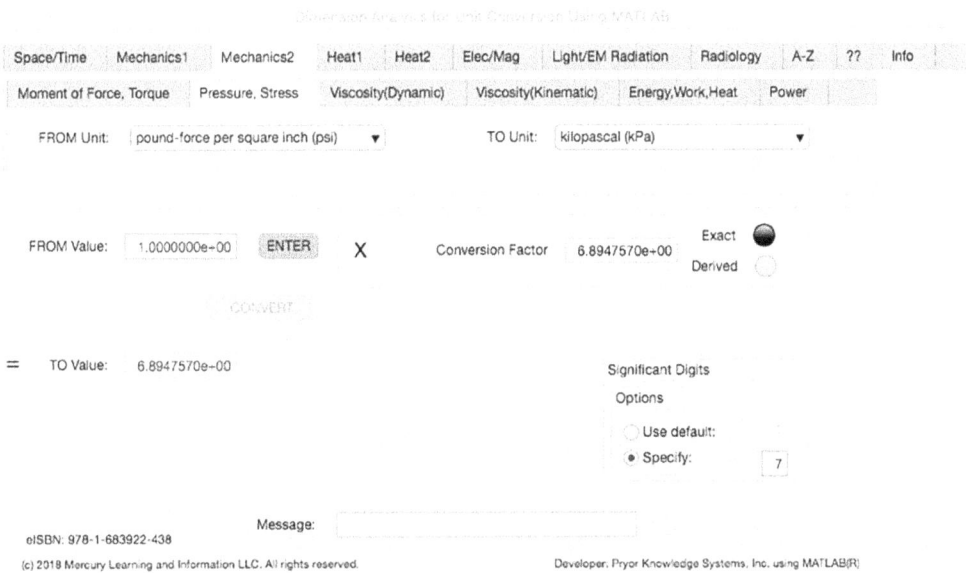

Space/Time	Mechanics1	Mechanics2	Heat1	Heat2	Elec/Mag	Light/EM Radiation	Radiology	A-Z	??	Info

Moment of Force, Torque	Pressure, Stress	Viscosity(Dynamic)	Viscosity(Kinematic)	Energy,Work,Heat	Power

FROM Unit: pound-force per square inch (psi) ▼   TO Unit: kilopascal (kPa) ▼

FROM Value: 1.0000000e+00   ENTER   X   Conversion Factor 6.8947570e+00   Exact ● / Derived ○

CONVERT

= TO Value: 6.8947570e+00

Significant Digits
Options
○ Use default:
● Specify: 7

Message: [_____]

oISBN: 978-1-683922-438
Developer: Pryor Knowledge Systems, Inc. using MATLAB(R)

FIGURE 3.3-2 Unit Converter Using MATLAB App - Mechanics2 Pressure, Stress Tab.

The second Minor Tab under the Mechanics2 Major Tab is the Pressure, Stress Tab (see Figure 3.3-2). This example demonstrates the conversion of pound-force per square inch to kilopascal. In this example, the amplitude of pound-force per square inch (psi) (lbf/in^2) is set equal to one (1.0). The conversion equation is shown in Example 3.3-2.

pound-force per square inch (psi) (=1.0) to kilopascal (kPa) Example 3.3-2:

kPa = lbf/in^2* in^2/m^2 * N/lbf * kPa/Pa
   = lbf/in^2* 1.550031E3 * 4.448222E0 * 1.0E-3
   = lbf/in^2* 6.894757E0

Where: kPa = the number of kilopascal calculated
lbf/in^2 (psi) = the number of lbf/in^2 (psi) to be converted (1.0)
   in^2/m^2 = the conversion factor from m^2 to in^2 (1.550031E3)
      N/lbf = the conversion factor from lbf to N (4.448222E0)
     kPa/Pa = the conversion factor from Pa to kPa (1.0E-3)

And:

   The resulting solution is:

   kPa = lbf/in^2* 6.894757E0= 6.894757E0 kilopascal   (3.3-2)

The Mechanics2 Pressure, Stress Tab allows the bidirectional selection of the following Pressure, Stress dimensions:

Major-Tab	Minor-Tab	Units on Pull-Down
Mechanics2		
	Pressure, Stress	
		megapascal (Mpa)
		standard atmosphere
		bar
		kilopascal (kPa)
		millibar
		pound-force per square inch (psi)
		kilopound-force per square inch
		pound-force per square foot
		inch of mercury (32 degF)
		foot of water (39.2 degF)
		inch of water (39.2 degF)
		millimeter of mercury (32 degF)
		torr (Torr)
		pascal (Pa)

Viscosity (Dynamic) Tab

The third Minor Tab under the Mechanics2 Major Tab is the Viscosity (Dynamic) Tab (see Figure 3.3-3). This example demonstrates the conversion of centipoise to millipascal second. In this example, the amplitude of centipoise (cP) is set equal to one (1.0). The conversion equation is shown in Example 3.3-3.

centipoise (cP) (=1.0) to millipascal second (mPa*s) Example 3.3-3:

$$mPa^*s = cP^* \ mPa^*s/cP = 1.0^* \ 1.0 = cP^* \ 1.0$$

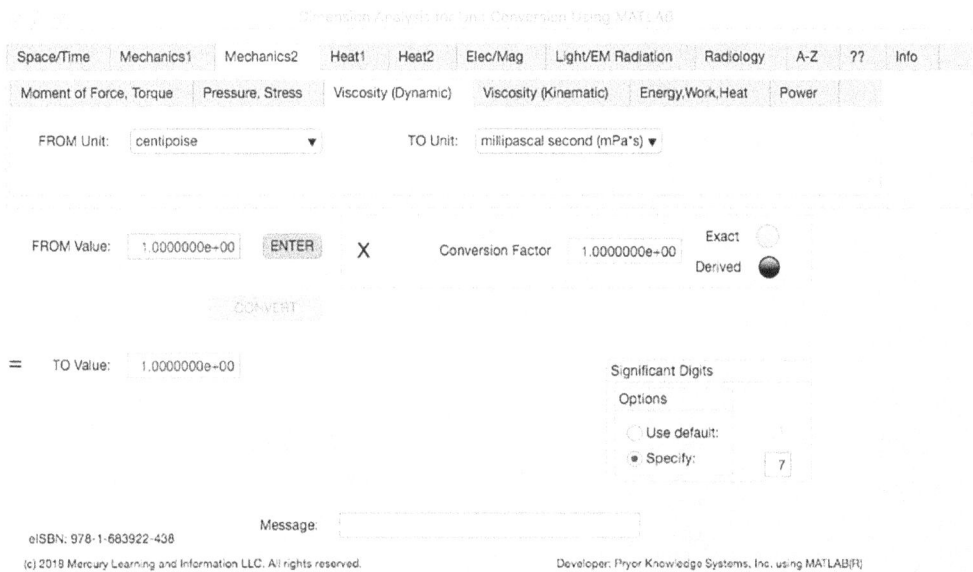

FIGURE 3.3-3 Unit Converter Using MATLAB App - Mechanics2 Viscosity (Dynamic) Tab.

Where: mPa*s = the number of millipascal calculated
    cP = the number of centipoise to be converted (1.0)
    mPa*s/cP = the conversion factor from centipoise to
        millipascal*second (1.0)

And:

The resulting solution is:

$$mPa^*s = cP^* \ mPa^*s/cP = 1.0^* \ 1.0 = 1.0 \ \text{millipascal second} \qquad (3.3-3)$$

The Mechanics2 Viscosity (Dynamic) Tab allows the bidirectional selection of the following Viscosity (Dynamic) dimensions:

Major-Tab	**Minor-Tab**	**Units on Pull-Down**
Mechanics2		
	Viscosity (Dynamic)	
		centipoise
		millipascal second (mPa*s)

Viscosity (Kinematic) Tab

The fourth Minor Tab under the Mechanics2 Major Tab is the Viscosity (Kinematic) Tab (see Figure 3.3-4). This example demonstrates the conversion of centistokes to square millimeter per second. In this example, the amplitude of centistokes (cSt) is set equal to one (1.0). The conversion equation is shown in Example 3.3-4.

centistokes (cSt) (=1.0) to square millimeter per second (mm^2/s) Example 3.3-4:

$$mm^2/s = cSt * (mm^2/s)/cSt = 1.0 * 1.0 = cSt * 1.0$$

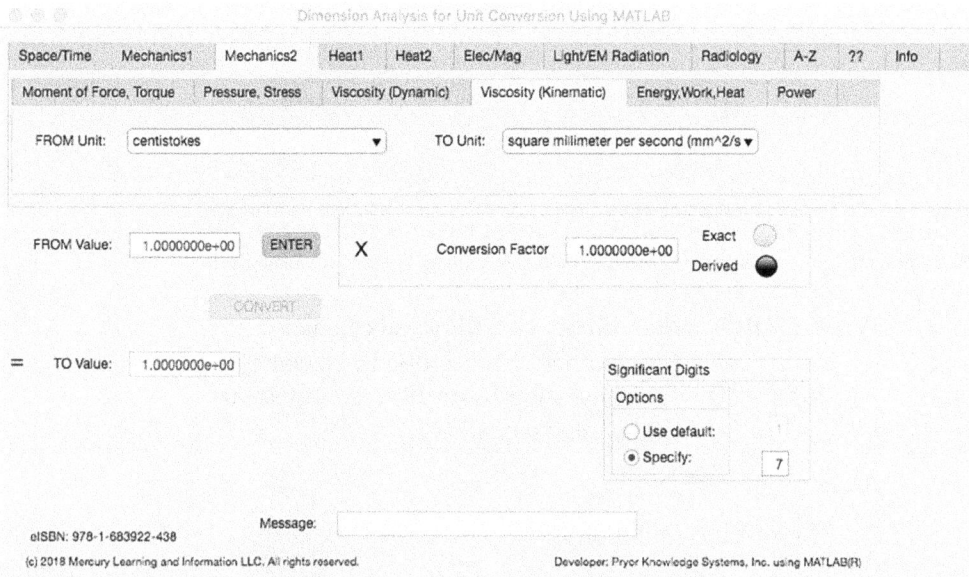

FIGURE 3.3-4 Unit Converter Using MATLAB App - Mechanics2 Viscosity (Kinematic) Tab.

Where: mm^2/s = the number of millimeter squared per second calculated
cSt = the number of centistoke to be converted (1.0)
(mm^2/s)/cSt = the conversion factor from centistoke to
mm^2 per second (1.0)

And:

The resulting solution is:

mm^2/s = cSt * (mm^2/s)/cSt = 1.0 * 1.0 = 1.0 millimeter
squared per second (3.3-4)

The Mechanics2 Viscosity (Kinematic) Tab allows the bidirectional selection of the following Viscosity (Kinematic) dimensions:

Major-Tab	Minor-Tab	Units on Pull-Down
Mechanics2		
	Viscosity (Kinematic)	
		centistokes
		square millimeter per second (mm^2/s)

Energy, Work, Heat Tab

The fifth Minor Tab under the Mechanics2 Major Tab is the Energy, Work, Heat Tab (see Figure 3.3-5). This example demonstrates the conversion of kilowatthour to joule. In this example, the amplitude of kilowatthour (kW * h) is set equal to one (1.0). The conversion equation is shown in Example 3.3-5.

kilowatthour (kW * h) (=1.0) to joule (J) Example 3.3-5:

J = kW * h * W/kW * min/h * s/min
= kW * h * 1.0E3 * 6.0E1 * 6.0E1
= kW * h * 3.6E6

Where: J = the number of joule calculated
kW * h = the number of kilowatthour to be converted (1.0)
W/kW = the conversion factor from kilowatt to watt (1.0E3)
min/h = the conversion factor from hour to minute (6.0E1)
s/min = the conversion factor from minute to second (6.0E1)

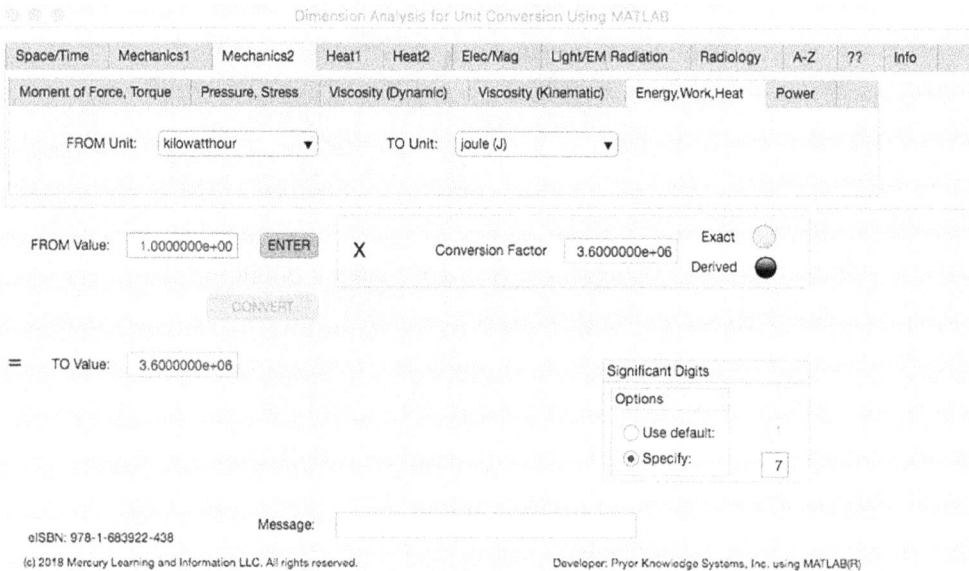

FIGURE 3.3-5 Unit Converter Using MATLAB App - Mechanics2 Energy, Work, Heat Tab.

And:

The resulting solution is:

$$J = 1.0 * 3.6E6 = 3.6E6 \text{ joule per kilowatthour} \qquad (3.3\text{-}5)$$

The Mechanics2 Energy, Work, Heat Tab allows the bidirectional selection of the following Energy, Work, Heat dimensions:

Major-Tab	Minor-Tab	Units on Pull-Down
Mechanics2		
	Energy, Work, Heat	
		kilowatthour
		megajoule (MJ)
		calorie (physics)
		kilojoule (kJ)
		calorie (nutrition) (kCal)
		joule (J)
		Btu
		therm (U.S.)
		horsepower hour
		foot pound-force

Power Tab

The sixth Minor Tab under the Mechanics2 Major Tab is the Power Tab (see Figure 3.3-6). This example demonstrates the conversion of Btu per second to Btu per hour. In this example, the amplitude of Btu per second (Btu/s) is set equal to one (1.0). The conversion equation is shown in Example 3.3-6.

Btu per second (Btu/s) (=1.0) to Btu per hour (Btu/h) 3.3-6:

$$\text{Btu/h} = \text{Btu/s} * \text{s/min} * \text{min/h}$$
$$= \text{Btu/s} * 6.0E1 * 6.0E1$$
$$= \text{Btu/s} * 3.6E3$$

Where: Btu/h = the number of Btu/h calculated
Btu/s = the number of Btu/s to be converted (1.0)
min/h = the conversion factor from hour to minute (6.0E1)
s/min = the conversion factor from minute to second (6.0E1)

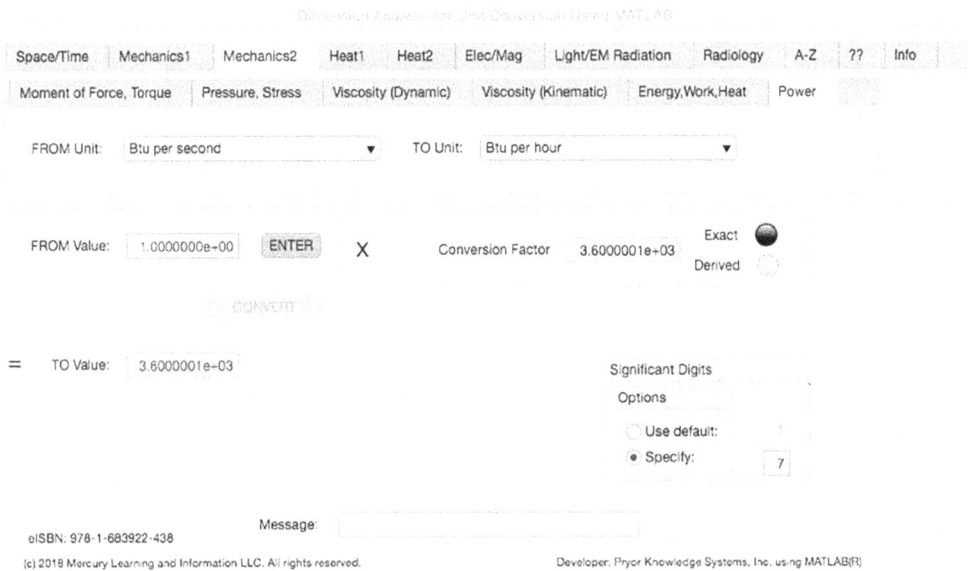

FIGURE 3.3-6 Unit Converter Using MATLAB App - Mechanics2 Power Tab.

And:

The resulting solution is:

Btu/h = 1.0 * 3.6E3 = 3.6E3 Btu per hour   (3.3-6)

The Mechanics2 Power Tab allows the bidirectional selection of the following Power dimensions:

Major-Tab	**Minor-Tab**	**Units on Pull-Down**
Mechanics2		
	Power	
		ton, refrigeration (12 000 Btu/h)
		kilowatt (kW)
		Btu per second
		Btu per hour
		watt (W)
		horsepower (550 ft-lbF/s)
		horsepower, electric
		foot pound-force per second (ft-lbF/s)

3.4 QUANTITIES OF HEAT1

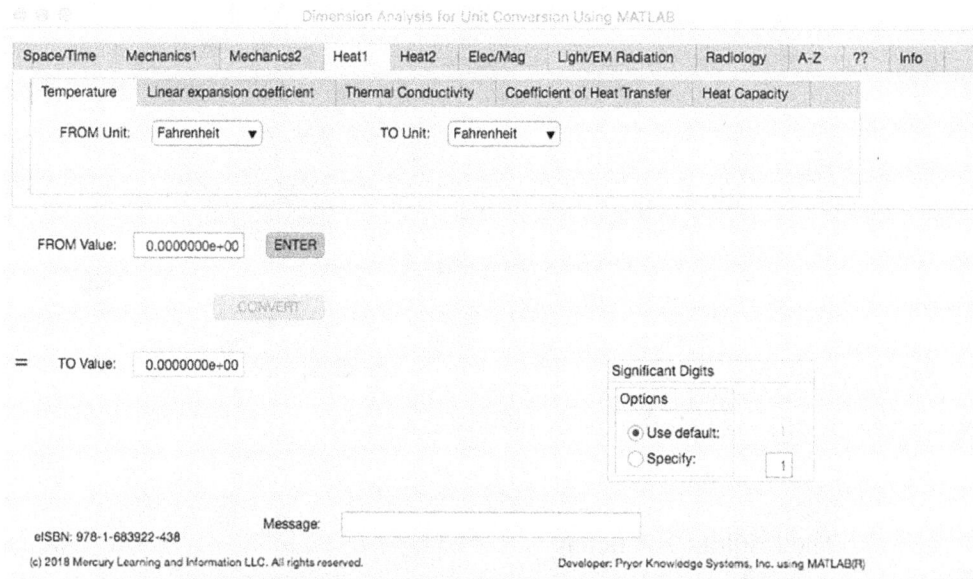

FIGURE 3.4-0 Unit Converter Using MATLAB App - Front Panel Heat1 Tab.

Figure 3.4-0 shows the Unit Converter MATLAB App Front Panel with the Heat1 Major Tab selected. Bidirectional matrices were created and installed within this app to allow the user the ability to easily select via pull-down menus all of the desired conversion pairs required. Each of the Heat1 Minor Tabs available in the App is explored herein.

Temperature Tab

The first Minor Tab under the Heat1 Major Tab is the Temperature Tab (see Figure 3.4-1). This example demonstrates the conversion of Fahrenheit to Celsius. In this example, the amplitude of Fahrenheit (°F) is set equal to fifty (50). The conversion equation is shown in Example 3.4-1.

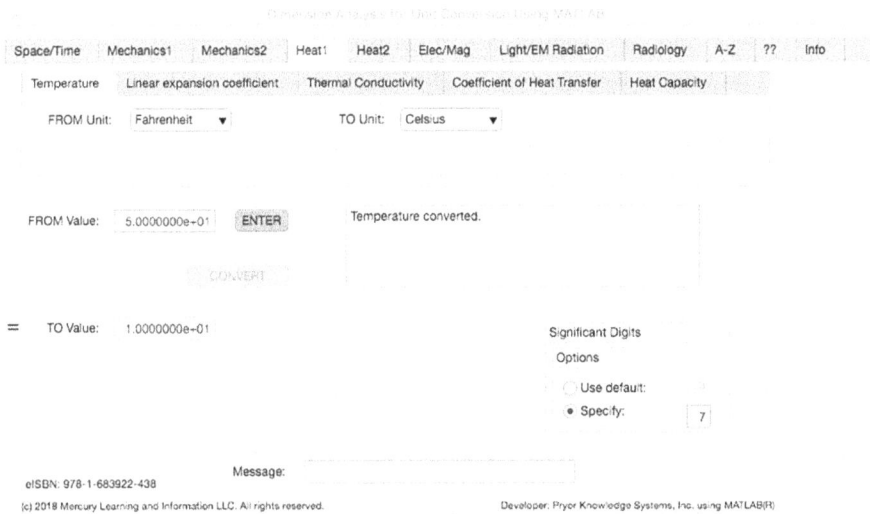

FIGURE 3.4-1 Unit Converter Using MATLAB App - Heat1 Temperature Tab.

Temperature (°F) (= 50.0) to Temperature (°C) Example 3.4-1:

$$°C = (°F-32)/1.8 = (50.0-32.0)/1.8 = 18.0/1.8 = 10$$

Where: °C = the Celsius temperature calculated
　　　°F = the Fahrenheit temperature being converted (50.0)
(°F-32)/1.8 = the conversion formula from Fahrenheit to Celsius

And:

The resulting solution is:

°F (50.0) = °C (10.0)　　　　　　　　　　　　　　　　　　　　(3.4-1)

The Heat1 Temperature Tab allows the bidirectional selection of the following Temperature dimensions:

Major-Tab	Minor-Tab	Units on Pull-Down
Heat1		
	Temperature	
		Fahrenheit
		Celsius
		Kelvin

Linear Expansion coefficient Tab

The second Minor Tab under the Heat1 Major Tab is the Linear Expansion coefficient Tab (see Figure 3.4-2). This example demonstrates the conversion of reciprocal Fahrenheit (°F) to reciprocal Celsius (°C). In this example, the amplitude of Fahrenheit (°F) is set equal to eighteen (18). The conversion equation is shown in Example 3.4-2.

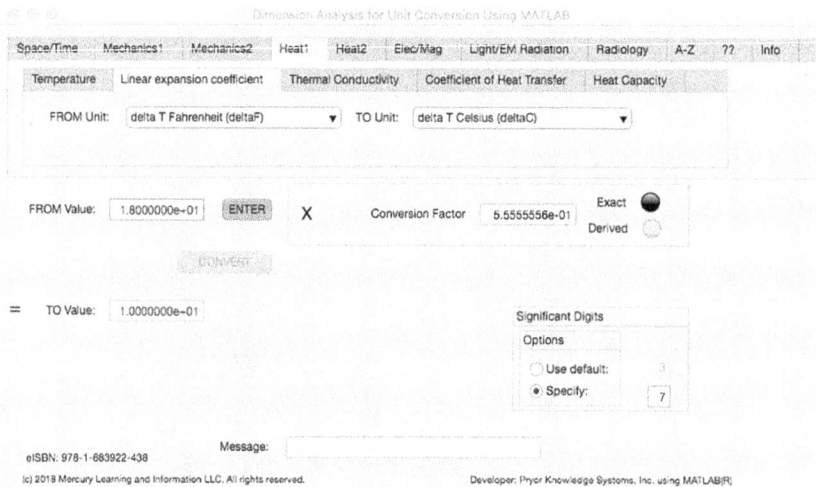

FIGURE 3.4-2 Unit Converter Using MATLAB App - Heat1 Linear Expansion coefficient Tab.

Temperature (°F) (=18.0) to Temperature (°C) Example 3.4-2:

$$\text{delta } (1/°C) = \text{delta } (1/°F)*1.8 = \text{delta } (1.8/18) = 0.1$$
$$\text{delta } (°C) = 1/0.1 = 10$$

Where: delta °C = the differential Celsius temperature calculated delta °F = the differential Fahrenheit temperature being converted (18.0)

And:

The resulting solution is:

$$\text{delta } °F \,(18.0) = \text{delta } °C \,(10.0) \tag{3.4-2}$$

The Heat1 Linear Expansion coefficient Tab allows the bidirectional selection of the following Linear Expansion coefficient dimensions:

Major-Tab Minor-Tab   **Units on Pull-Down**
Heat1
                Linear expansion coefficient

delta T Fahrenheit (delta °F)
delta T kelvin (delta K)
delta T Celsius (delta °C)

Thermal Conductivity Tab

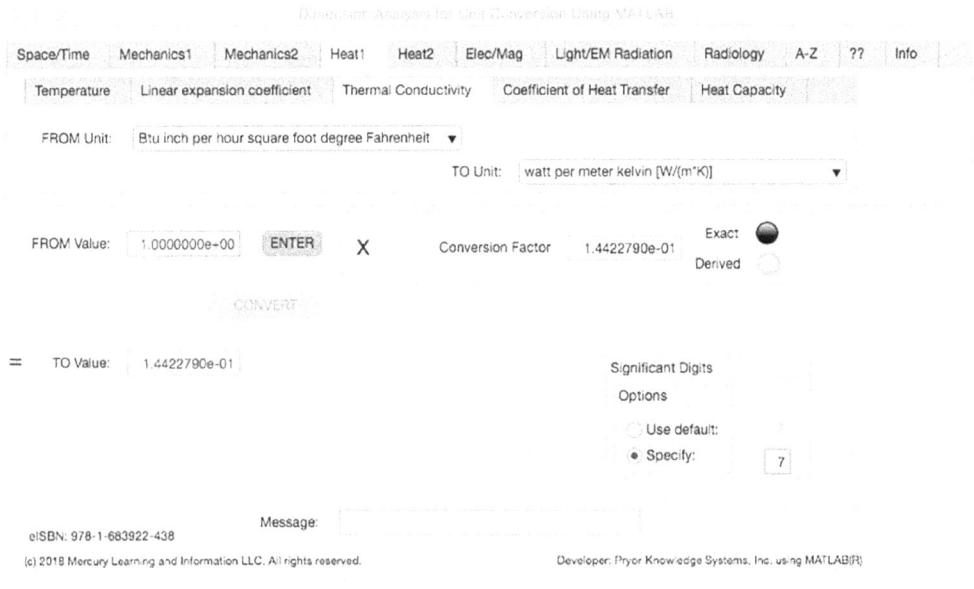

| Space/Time | Mechanics1 | Mechanics2 | Heat1 | Heat2 | Elec/Mag | Light/EM Radiation | Radiology | A-Z | ?? | Info |

| Temperature | Linear expansion coefficient | Thermal Conductivity | Coefficient of Heat Transfer | Heat Capacity |

FROM Unit: Btu inch per hour square foot degree Fahrenheit ▼

TO Unit: watt per meter kelvin [W/(m*K)] ▼

FROM Value: 1.0000000e+00 ENTER X Conversion Factor 1.4422790e-01 Exact ⬤ Derived ○

CONVERT

= TO Value: 1.4422790e-01

Significant Digits
Options
○ Use default:
● Specify: 7

Message:

FIGURE 3.4-3 Unit Converter Using MATLAB App - Heat1 Thermal Conductivity Tab.

The third Minor Tab under the Heat1 Major Tab is the Thermal Conductivity Tab (see Figure 3.4-3). This example demonstrates the conversion of Btu inch per hour square foot degree Fahrenheit to watt per meter kelvin. In this example, the amplitude of Btu inch per hour square foot degree Fahrenheit (Btu * in/(h*ft^2*F)) is set equal to one (1.0). The conversion equation is shown in Example 3.4-3.

Btu ° in/(h°ft^2°F) (=1.0) to W/(m°K) Example 3.4-3:

$$W/(m°K) = Btu ° in/(h°ft^2°F) ° h/min ° min/s ° m/in ° ft^2/m^2 ° F/K$$
$$= 1.0 ° 1.055056E3 ° 1/60 ° 1/60 ° 2.54E-2 ° 1/9.290304E-2 ° 1.8$$
$$= Btu ° in/(h°ft^2°F) ° 1.4422791E-1$$

Where: W/(m°K) = the SI Thermal Conductivity calculated
   Btu as joules = W °s ° 1.055056E3
Btu ° in/(h°ft^2°F) = Thermal Conductivity being converted (1.0)
   h/min = number of hours per minute (1/60)
   min/s = number of minutes per second (1/60)
   m/in = number of meter per inch (2.54E-2)
   ft^2/m^2 = number of square meter per square foot (1/9.290304E-2)
   F/K = number of degree Fahrenheit per degree kelvin (1.8)

And:

The resulting solution is:

W/(m°K) = 1.0 ° 1.4422791E-1 = 1.4422791E-1 watt
per meter kelvin   (3.4-3)

The Heat1 Thermal Conductivity Tab allows the bidirectional selection of the following Thermal Conductivity dimensions:

Major-Tab	**Minor-Tab**	**Units on Pull-Down**
Heat1		
	Thermal Conductivity	
		Btu inch per hour square foot degree Fahrenheit
		watt per meter kelvin [W/(m°K)]

Coefficient of Heat Transfer Tab

The fourth Minor Tab under the Heat1 Major Tab is the Coefficient of Heat Transfer Tab (see Figure 3.4-4). This example demonstrates the conversion of Btu inch per hour square foot degree Fahrenheit (Btu ° in/(h°ft^2°F)) to watt per square meter kelvin (W/(m^2°K)). In this example, the amplitude of Btu inch per hour square foot degree Fahrenheit (Btu ° in/(h°ft^2°F)) is set equal to one (1.0). The conversion equation is shown in Example 3.4-4.

Btu * in/(h°ft^2°F) (=1.0) to W/(m^2°K) Example 3.4-4:

$$
\begin{aligned}
W/(m^2°K) &= Btu * 1/(h°ft^2°F) * h/min * min/s * ft^2/m^2 * F/K \\
&= 1.0 * 1.055056E3 * 1/60 * 1/60 * 1/9.290304E\text{-}2 * 1.8 \\
&= Btu * 1/(h°ft^2°F) * 5.678264E0
\end{aligned}
$$

Space/Time	Mechanics1	Mechanics2	Heat1	Heat2	Elec/Mag	Light/EM Radiation	Radiology	A-Z	??	Info

Temperature	Linear expansion coefficient	Thermal Conductivity	Coefficient of Heat Transfer	Heat Capacity

FROM Unit: Btu per hour square foot degree Fahrenheit ▼   TO Unit: watt per square meter [W/(m^2*K)]   ▼

FROM Value: 1.0000000e+00   **ENTER**   **X**   Conversion Factor   5.6782630e+00   Exact ⬤ / Derived ◯

CONVERT

= TO Value: 5.6782630e-00

Significant Digits
Options
◯ Use default:
⦿ Specify: [7]

Message:

oISBN: 978-1-683922-438
Developer: Pryor Knowledge Systems, Inc. using MATLAB(R)

FIGURE 3.4-4 Unit Converter Using MATLAB App - Heat1 Coefficient of Heat Transfer Tab.

Where: W/(m^2°K) = the SI Coefficient of Heat Transfer calculated
Btu as joules = W °s * 1.055056E3
Btu * 1/(h°ft^2°F) = Coefficient of Heat Transfer being converted (1.0)
h/min = number of hours per minute (1/60)
min/s = number of minutes per second (1/60)
ft^2/m^2 = number of square meter per square foot
(1/9.290304E-2)
F/K = number of degree Fahrenheit per degree kelvin (1.8)

And:

The resulting solution is:

$$
\begin{aligned}
W/(m^2°K) &= Btu * 1/(h°ft^2°F) * 5.678264E0 = \\
1.0 &* 5.678264E0 = 5.678264E0 \text{ watts per square meter}
\end{aligned}
\tag{3.4-4}
$$

The Heat1 Coefficient of Heat Transfer Tab allows the bidirectional selection of the following Coefficient of Heat Transfer dimensions:

Major-Tab Minor-Tab   Units on Pull-Down

Heat1

Coefficient of
Heat Transfer

Btu per hour square foot degree Fahrenheit

watts per square meter kelvin [W/(m^2*K)]

Heat Capacity Tab

The fifth Minor Tab under the Heat1 Major Tab is the Heat Capacity Tab (see Figure 3.4-5). This example demonstrates the conversion of Btu per degree Fahrenheit (Btu/ °F) to kilojoule per kelvin (kJ/K). In this example, the amplitude of Btu per degree Fahrenheit (Btu/F)) is set equal to one (1.0). The conversion equation is shown in Example 3.4-5.

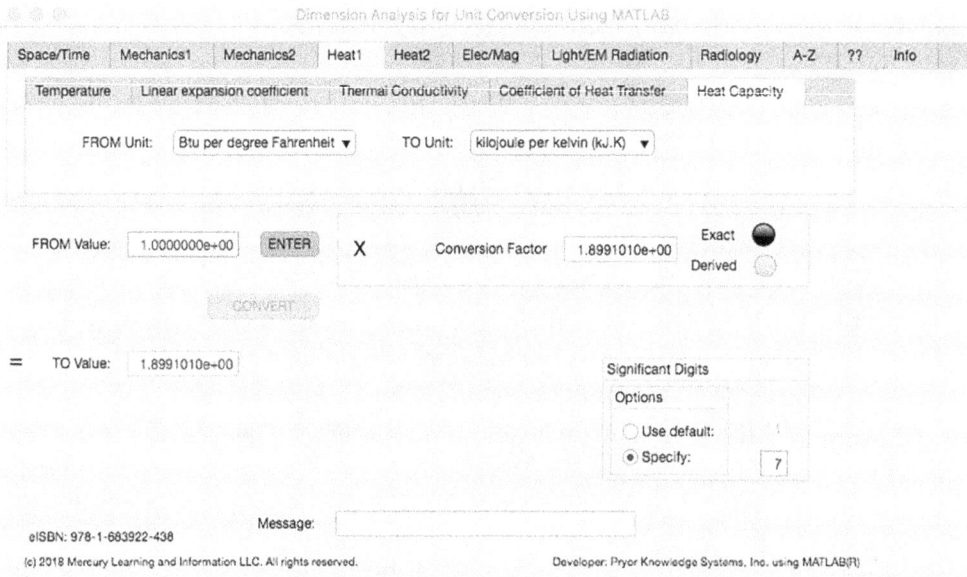

FIGURE 3.4-5 Unit Converter Using MATLAB App - Heat1 Heat Capacity Tab.

Btu/ °F (=1.0) to kJ/K Example 3.4-5:

$$kJ/K = Btu/ °F * kJ/J * °F/K$$
$$= Btu/ °F * 1.055056E3 * 1.0E\text{-}3 * 1.8$$
$$= Btu/ °F * 1.899101E0$$

Where: kJ/K = the SI Coefficient of Heat Capacity calculated
Btu as joules = W $^\circ$s $^\circ$ 1.055056E3
      Btu/ °F = Coefficient of Heat Capacity being converted (1.0)
        °F/K = number of degree Fahrenheit per degree kelvin (1.8)
        kJ/J = 1.0E-3

And:

The resulting solution is:

kJ/K = 1.0 $^\circ$ 1.899101E0 = 1.899101E0 kilojoules per kelvin   (3.4-5)

The Heat1 Heat Capacity Tab allows the bidirectional selection of the following Heat Capacity dimensions:

Major-Tab	Minor-Tab	Units on Pull-Down
Heat1		
	Heat Capacity	
		Btu per degree Fahrenheit
		kilojoule per kelvin (kJ/K)

3.5 QUANTITIES OF HEAT2

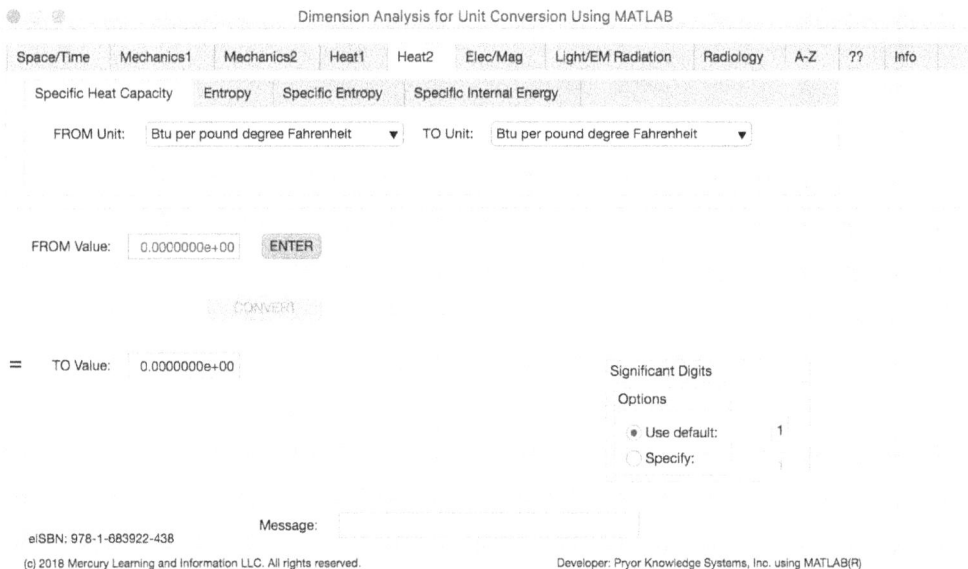

FIGURE 3.5-0 Unit Converter Using MATLAB App - Front Panel Heat2 Tab.

Figure 3.5-0 shows the Unit Converter MATLAB App Front Panel with the Heat2 Tab selected. Bidirectional matrices were created and installed within this app to allow the user the ability to easily select via pull-down menus all of the desired conversion pairs required. Each of the Tabs available in the App is explored herein.

Specific Heat Capacity Tab

The first Minor Tab under the Heat2 Major Tab is the Specific heat capacity Tab (see Figure 3.5-1). This example shows the conversion of Btu per pound degree Fahrenheit (Btu/(lb * °F) to Kilojoule per kilogram kelvin [kJ/(kg * K)]. In this example, the amplitude of Btu per pound degree Fahrenheit is set equal to one (1.0). The conversion equation is shown in Example 3.5-1.

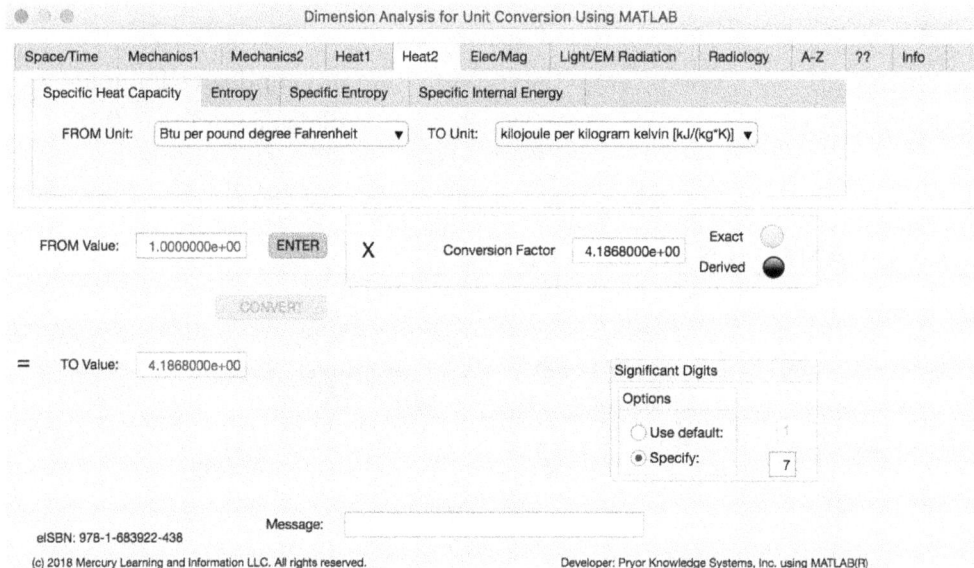

FIGURE 3.5-1 Unit Converter Using MATLAB App - Heat2 Specific Heat Capacity Tab.

Btu/(lb* °F) (=1.0) to kJ/(kg * K) Example 3.5-1:

$$kJ/(kg * K) = Btu/(lb * °F) * J/Btu * kJ/J * lb/kg * F/K$$
$$= Btu/(lb* °F) * 1.055056E3 * 1/1000$$
$$* 1/4.535924E\text{-}1 * 1/5.55556E\text{-}1$$
$$= BTU/(lb * F) * 4.1868E0$$

Where: kJ/(kg ° K) = the number of kilojoule per kilogram kelvin calculated

Btu/(lb°F) = the number of Btu per pound degree Fahrenheit to be converted (1.0)

J/BTU = the conversion factor from Btu to joule (1.055056E3)

kJ/J = the conversion factor from joule to kilojoule (1/1000)

lb/kg = the conversion factor from kilogram to pound (1/4.535924E-1)

F/K = the conversion factor from kelvin degree interval to Fahrenheit degree interval (1/5.55556E-1)

And:

The resulting solution is:

kJ/(kg ° K) = 1.0 ° 4.1868E+0 = 4.1868E+0 kilojoule per kilogram kelvin (3.5-1)

The Heat2 Specific heat capacity Tab allows the bidirectional selection of the following Specific heat capacity dimensions:

Major-Tab	**Minor-Tab**	**Units on Pull-Down**
Heat2		
	Specific Heat Capacity	
		Btu per pound degree Fahrenheit
		kilojoule per kilogram kelvin (KJ/(kg ° K))

Entropy Tab

The second Minor Tab under the Heat2 Major Tab is the Entropy Tab (see Figure 3.5-2). This example shows the conversion of Btu per degree Rankine (Btu/R) to kilojoule per kelvin (kJ/K). In this example, the amplitude of Btu per degree Rankine is set equal to one (1.0). The conversion equation is shown in Example 3.5-2.

Btu/R (=1.0) to kJ/K Example 3.5-2:

kJ/(K) = Btu/R ° J/Btu ° kJ/J ° R/K
= Btu/R ° 1.055056E3 ° 1/1000 ° 1/5.55556E-1
= Btu/R ° 1.899101E0

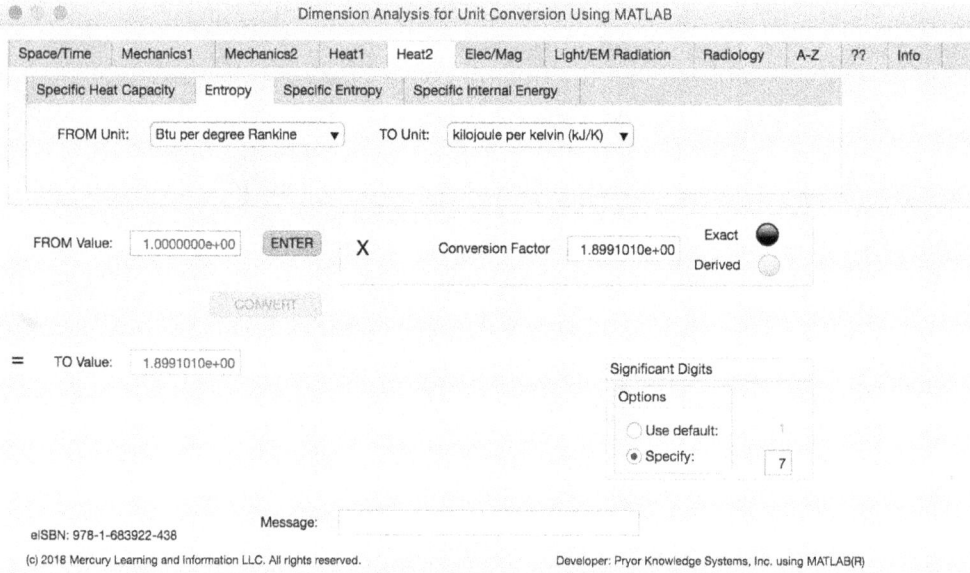

FIGURE 3.5-2 Unit Converter Using MATLAB App - Heat2 Entropy Tab.

Where: kJ/K = the number of kilojoule per kelvin calculated
Btu/R = the number of Btu per degree Rankine to be converted (1.0)
J/Btu = the conversion factor from Btu to joule (1.055056E3)
kJ/J = the conversion factor from joule to kilojoule (1/1000)
R/K = the conversion factor from kelvin degree interval to Rankine
degree interval (1/5.55556E-1)

And:

The resulting solution is:

kJ/K = 1.0 * 1.899101E0 = 1.899101E0 kilojoule per kelvin   (3.5-2)

The Heat2 Entropy Tab allows the bidirectional selection of the following
entropy dimensions:

Major-Tab	Minor-Tab	Units on Pull-Down
Heat2		
	Entropy	
		Btu per degree Rankine (BTU/R)
		kilojoule per kelvin (kJ/K)

Specific Entropy

The third Minor Tab under the Heat2 Major Tab is the Specific entropy Tab (see Figure 3.5-3). This example shows the conversion of Btu per pound degree Rankine [Btu/(lb°R)] to kilojoule per kilogram kelvin [kJ/(kg°K)]. In this example, the amplitude of Btu per pound degree Rankine [Btu/(lb°R)] is set equal to one (1.0). The conversion equation is shown in Example 3.5-3.

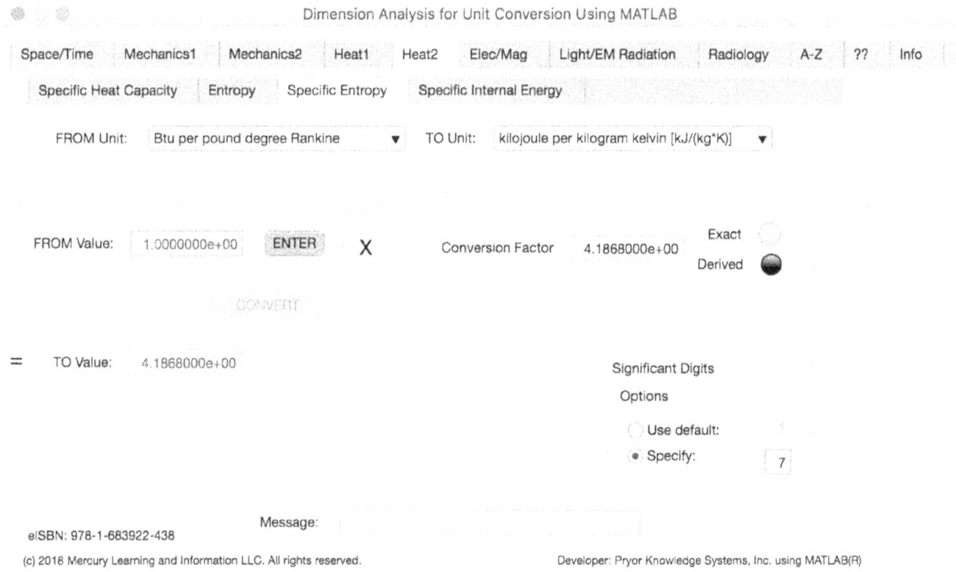

FIGURE 3.5-3 Unit Converter Using MATLAB App - Heat2 Specific Entropy Tab.

Btu/(lb°R) (=1.0) to kJ/(kg°K) Example 3.5-3:

$$kJ/(kg°K) = Btu/(lb°R)°J/Btu°kJ/J°lb/kg°R/K$$
$$= Btu/(lb°R)°1.055056E3°1/1000$$
$$°1/4.535924E-1°1/5.55556E-1$$
$$= Btu/(lb°R)°4.1868E0$$

Where: kJ/(kg°K) = the number of kilojoule per kilogram kelvin calculated
Btu/(lb°R) = the number of Btu per pound degree Fahrenheit to be converted (1.0)
J/Btu = the conversion factor from Btu to joule (1.055056E3)
kJ/J = the conversion factor from joule to kilojoule (1/1000)

lb/kg = the conversion factor from kilogram to pound
(1/4.535924E-1)

R/K = the conversion factor from kelvin degree interval to
Fahrenheit degree interval (1/5.55556E-1)

And:

The resulting solution is:

kJ/(kg°K) = 1.0°4.1868E+0 = 4.1868E+0 kilojoule
per kilogram kelvin   (3.5-3)

The Heat2 Specific Entropy Tab allows the bidirectional selection of the following Specific Entropy dimensions:

Major-Tab **Minor-Tab**   **Units on Pull-Down**

Heat2

Specific Entropy

Btu per pound degree Rankine [Btu/(lb ° R)]

kilojoule per kilogram kelvin ([kJ/(kg°K)])

Specific Internal Energy Tab

The fourth Minor Tab under the Heat2 Major Tab is the Specific Internal Energy Tab (see Figure 3.5-4). This example shows the conversion of Btu per pound (Btu/lb) to kilojoule per kilogram (kJ/kg). In this example, the amplitude of Btu per pound is set equal to one (1.0). The conversion equation is shown in Example 3.5-4.

Btu per pound to kilojoule per kilogram (kJ/kg) Example 3.5-4:

kJ/kg = Btu/lb° J/Btu ° kJ/J ° lb/kg
= Btu/lb ° 1.055056E3 ° 1/1000 ° 1/4.535924E-1
= Btu/lb ° 2.326E0

Where: kJ/kg = the number of kilojoule per kilogram calculated
Btu/lb = the number of Btu per pound to be converted (1.0)
J/Btu = the conversion factor from Btu to joule (1.055056E3)

kJ/J = the conversion factor from joule to kilojoule (1/1000)

lb/kg = the conversion factor from kilogram to pound

(1/4.535924E-1)

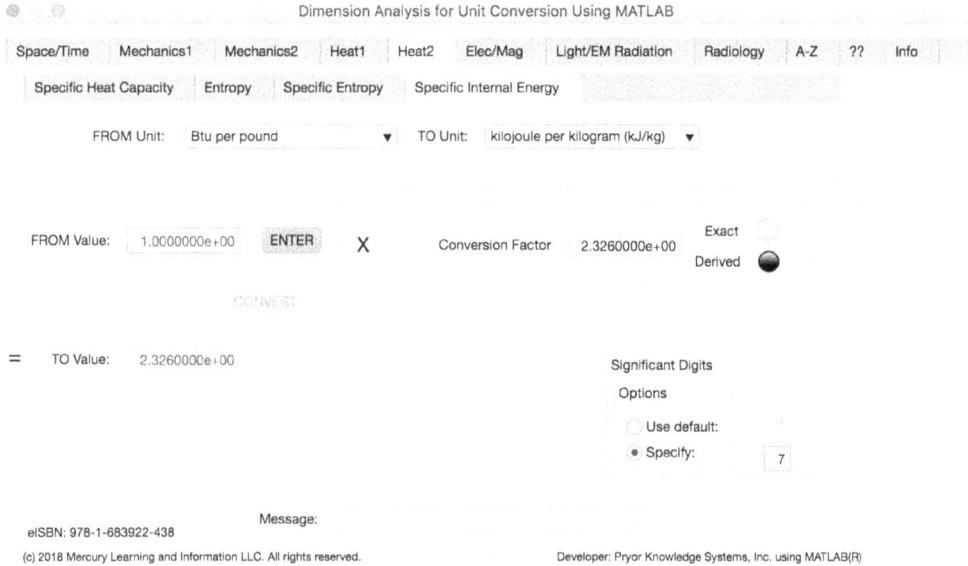

FIGURE 3.5-4 Unit Converter Using MATLAB App - Heat2 Specific Internal Energy Tab.

And:

The resulting solution is:

kJ/kg = 1.0 * 2.326E0 = 2.326E0 kilojoule per kilogram   (3.5-4)

The Heat2 Specific Internal Energy Tab allows the bidirectional selection of the following specific internal energy dimensions:

Major-Tab	**Minor-Tab**	**Units on Pull-Down**
Heat2		
	Specific Internal Energy	
		Btu per pound (Btu/lb)
		kilojoule per kilogram (kJ/kg)

3.6 QUANTITIES OF ELECTRICITY AND MAGNETISM

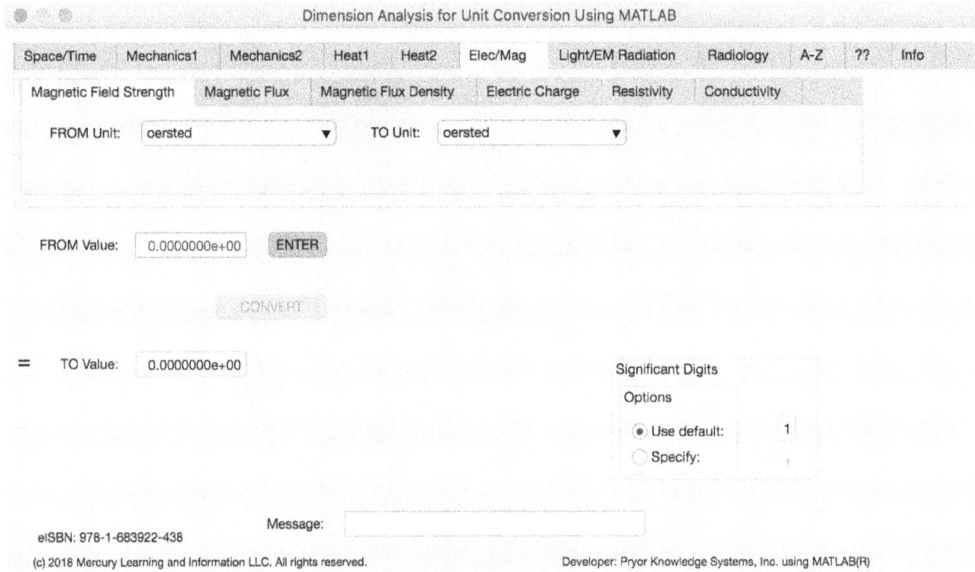

FIGURE 3.6-0 Unit Converter Using MATLAB App - Front Panel Elec/Mag Tab.

Figure 3.6-0 shows the Unit Converter MATLAB App Front Panel with the Electricity and Magnetism (ElecMag) Tab selected. Bidirectional matrices were created and installed within this app to allow the user the ability to easily select via pull-down menus all of the desired conversion pairs required. Each of the Tabs available in the App is explored herein.

Magnetic Field Strength Tab

The first Minor Tab under the Elec/Mag Major Tab is the Magnetic Field Strength Tab (see Figure 3.6-1). This example demonstrates the conversion of oersted to ampere/meter (A/m). In this example, the amplitude of oersted is set equal to one (1.0). The conversion equation is shown in Example 3.6-1 {37}.

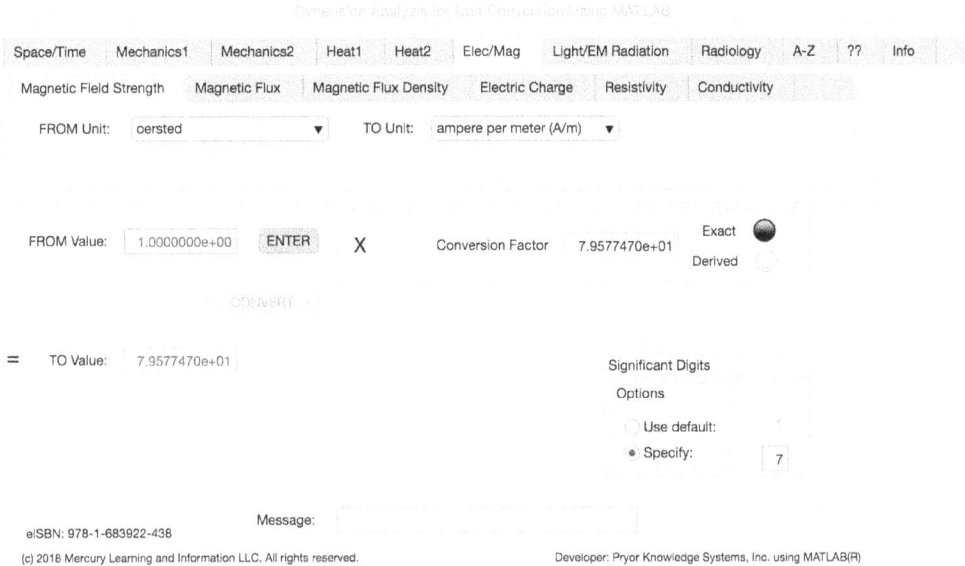

FIGURE 3.6-1 Unit Converter Using MATLAB App - Elec/Mag Magnetic Field Strength Tab.

oersted (=1.0) to ampere/meter (A/m) Example 3.6-1:

A/m = oersted * (A/m)/oersted = 1.0 * 1000/4π = 1.0 * 79.57747

Where: A/m = the number of amperes per meter calculated
oersted = the number of oersted to be converted (1.0)
(A/m)/oersted = the conversion factor from oersted to ampere/meter
(79.57747)

And:

The resulting solution is:

A/m = 1.0* 7.957747E+1 = 7.957747E+1 ampere per meter   (3.6-1)

The Elec/Mag Magnetic Field Strength Tab allows the bidirectional selection of the following Magnetic Field Strength dimensions:

Major-Tab	Minor-Tab	Units on Pull-Down
Elec/Mag		
	Magnetic Field Strength	
		oersted
		ampere per meter (A/m)

Magnetic Flux Tab

The second Minor Tab under the Elec/Mag Major Tab is the Magnetic Flux Tab (see Figure 3.6-2). This example shows the conversion of maxwell to nanoweber. In this example, the amplitude of maxwell is set equal to one (1.0). The conversion equation is shown in Example 3.6-2.

maxwell (Mx) (=1.0) to nanoweber (nWb) Example 3.6-2:

$$nWb = maxwell * Wb/Mx * nWb/Wb$$
$$= maxwell * 10^{-8} * 10^{9}$$
$$= maxwell * 10$$
$$= 1.0E+1 \text{ nWb}$$

Where: nWb = the number of nanoweber calculated
Mx = the starting number of maxwell (1.0)
Wb/Mx = the conversion factor from maxwell to weber (10^{-8})
nWb/Wb = the conversion factor from weber to nanoweber (10^{9})

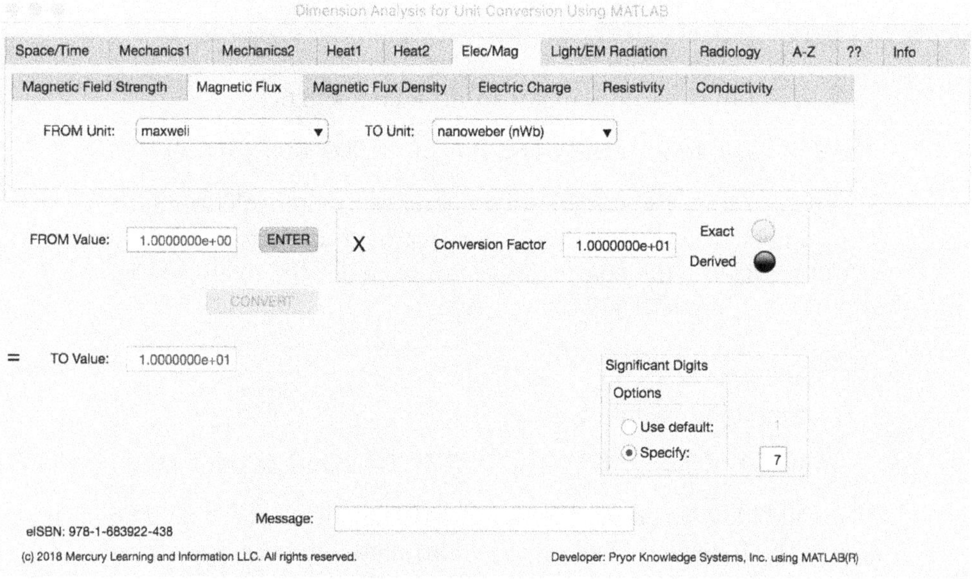

FIGURE 3.6-2 Unit Converter Using MATLAB App - Elec/Mag Magnetic Flux Tab.

And:

The resulting solution is:

nWb = 1.0 * 10 = 10 nanoweber   (3.6-2)

The Elec/Mag Magnetic Flux Tab allows the bidirectional selection of the following magnetic flux dimensions:

Major-Tab	**Minor-Tab**	**Units on Pull-Down**
Elec/Mag		
	Magnetic Flux	
		maxwell (Mx)
		nanoweber (nWb)

Magnetic Flux Density Tab

The third Minor Tab under the Elec/Mag Major Tab is the Magnetic Flux Density Tab (see Figure 3.6-3). This example shows the conversion of gauss to millitesla. In this example, the amplitude of gauss is set equal to one (1.0). The conversion equation is shown in Example 3.6-3.

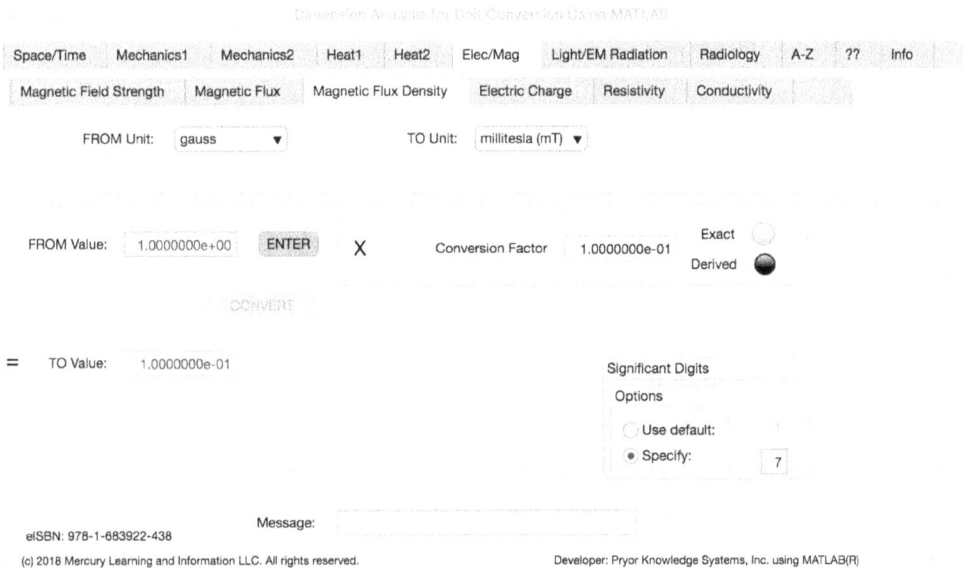

FIGURE 3.6-3 Unit Converter Using MATLAB App - Elec/Mag Magnetic Flux Density Tab.

gauss (G) to millitesla (mT) Example 3.6-3:

$$mT = G * T/ G * mT/T = G * 10^{-4} * 10^3 = gauss * 1.0E\text{-}1$$

Where: mT = the number of millitesla calculated
G = the number of gauss to be converted (1.0)

$\quad$ T/G = the conversion factor from gauss to tesla (10^{-4})

$\quad$ mT/T = the conversion factor from tesla to millitesla (10^3)

And:

$\quad$ The resulting solution is:

$\quad$ mT = 1.0 $\ast$ 1.0E-1 = 1.0E-1 millitesla $\hfill$ (3.6-3)

The Elec/Mag Magnetic Flux Density Tab allows the bidirectional selection of the following Magnetic Flux Density dimensions:

Major-Tab	**Minor-Tab**	**Units on Pull-Down**
Elec/Mag		
	Magnetic Flux Density	
		gauss (G)
		millitesla (mT)

Electric Charge Tab

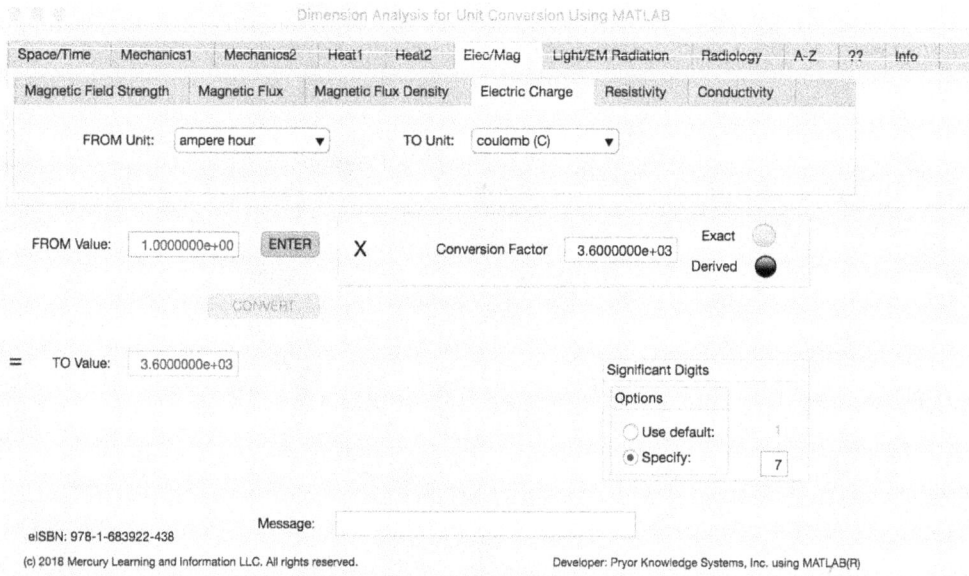

FIGURE 3.6-4 Unit Converter Using MATLAB App - Elec/Mag Electric Charge Tab.

The fourth Minor Tab under the Elec/Mag Major Tab is the Electric Charge Tab (see Figure 3.6-4). This example shows the conversion of ampere hour to coulomb. In this example, the amplitude of ampere hour is set equal to one (1.0). The conversion equation is shown in Example 3.6-4.

ampere hour (A*h) (=1.0) to coulomb (C) Example 3.6-4:

$$C = A*h * (C/s)(1/A) * min/h * s/min$$
$$= A*h * 1 * 60 * 60$$
$$= A*h * 3600 = A*h * 3.6E+3$$

Where: C = the number of coulomb calculated
   A*h = the number of ampere hour to be converted (1.0)
(C/s)(1/A) = the conversion factor from ampere to coulomb per second (1)
   min/h = the conversion factor from hours to minutes (60)
   s/min = the conversion factor from minutes to seconds (60)

And:

The resulting solution is:

$$C = 1.0 * 3.6E+3 = 3.6E+3 \text{ coulomb} \tag{3.6-4}$$

The Elec/Mag Electric Charge Tab allows the bidirectional selection of the following Electric Charge dimensions:

Major-Tab	**Minor-Tab**	**Units on Pull-Down**
Elec/Mag		
	Electric Charge	
		ampere hour (A*h)
		coulomb (C)

Resistivity Tab

The fifth Minor Tab under the Elec/Mag Major Tab is the Resistivity Tab (see Figure 3.6-5). This example shows the conversion of ohm circular mil per foot to Nanoohm meter (nOhm * m). In this example, the amplitude of ohm circular mil per foot is set equal to one (1.0). The conversion equation is shown in Example 3.6-5.

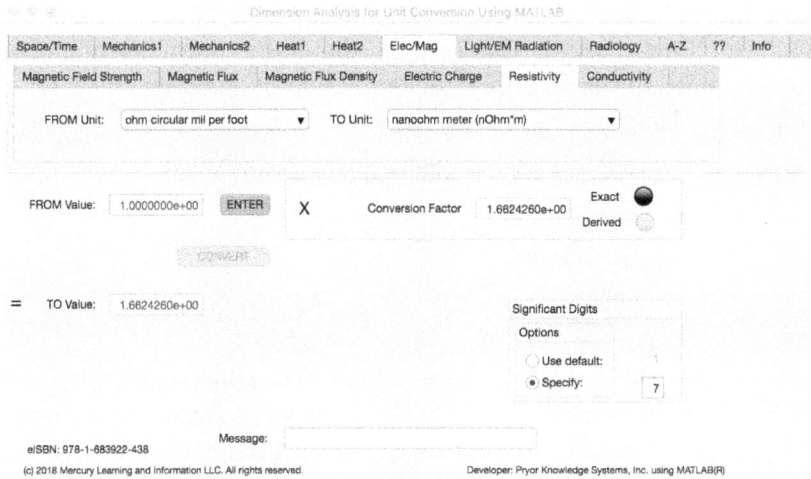

FIGURE 3.6-5 Unit Converter Using MATLAB App - Elec/Mag Resistivity Tab.

ohm circular mil per foot (Ω*cmil/ft) (=1.0) to nanoohm meter (nΩ * m) Example 3.6-5:

$$n\Omega^* m = \Omega^*cmil/ft * n\Omega/\Omega * m^2/cmil * ft/m$$
$$= \Omega^*cmil/ft * 1.0E+9 * 5.067075E\text{-}10 * 1/0.3048$$
$$= \Omega^*cmil/ft * 1.662426E+0$$

Where: nΩ * m = the number of nanoohm meter calculated

    Ω*cmil/ft = the number of ohm circular mil per foot to be converted (1.0)

    nΩ/Ω = the conversion factor from ohm to nanoohm (1.0E+9)

    m^2/cmil = the conversion factor from circular mil to meter squared (5.067075E-10)

    ft/m = the conversion factor from meter to foot (1/0.3048)

And:

The resulting solution is:

$$n\Omega * m = 1.0 * 1.662426E+0 = 1.662426E+0 \text{ nanoohm meter} \quad (3.6\text{-}5)$$

The Elec/Mag Resistivity Tab allows the bidirectional selection of the following Resistivity dimensions:

Major-Tab	Minor-Tab	Units on Pull-Down
Elec/Mag		
	Resistivity	
		ohm circular mil per foot
		nanoohm meter (nΩ * m)

Conductivity Tab

The sixth Minor Tab under the Elec/Mag Major Tab is the Conductivity Tab (see Figure 3.6-6). This example shows the conversion of mho per centimeter (mho/cm) to siemens/meter (S/m). In this example, the amplitude of mho per centimeter (mho/cm) is set equal to one (1.0). The conversion equation is shown in Example 3.6-6.

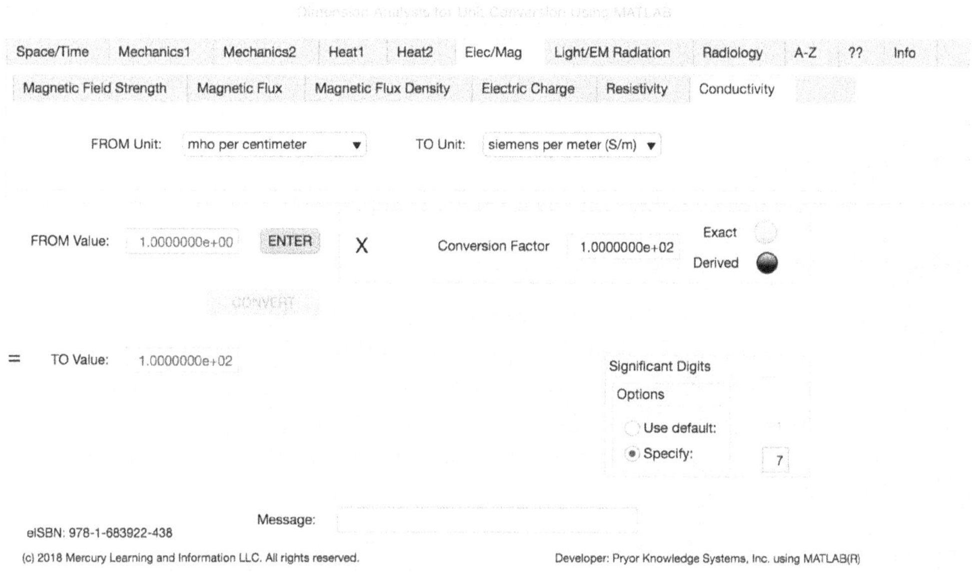

FIGURE 3.6-6 Unit Converter Using MATLAB App - Elec/Mag Conductivity Tab.

mho per centimeter (mho/cm) (=1.0) to siemens per meter (S/m) Example 3.6-6:

$$S/m = mho/cm * cm/m * S/mho$$
$$= mho/cm * 100 * 1.0$$
$$= mho/cm * 100 = mho/cm * 1.0E+2$$

Where: S/m = the number of siemens per meter calculated
mho/cm = the number of mho per centimeter to be converted (1.0)
cm/m = the conversion factor from meter to centimeter = 100
S/mho = the conversion factor from mho to siemens (1.0)

And:

The resulting solution is:

$$S/m = 1.0 * 100 = 100 \text{ siemens per meter} \tag{3.6-6}$$

The Elec/Mag Conductivity Tab allows the bidirectional selection of the following conductivity dimensions:

Major-Tab	**Minor-Tab**	**Units on Pull-Down**
Elec/Mag		
	Conductivity	
		mho per centimeter (mho/cm)
		siemens per meter (S/m)

3.7   QUANTITIES OF LIGHT AND RELATED ELECTROMAGNETIC RADIATION (LIGHT/EM RADIATION)

Figure 3.7-0 shows the Unit Converter MATLAB App Front Panel with the Light and Related Electromagnetic Radiation (Light/EM Radiation) Tab selected. Bidirectional matrices were created and installed within this app to allow the user the ability to easily select via pull-down menus all of the desired conversion pairs required. Each of the Tabs available in the App is explored herein.

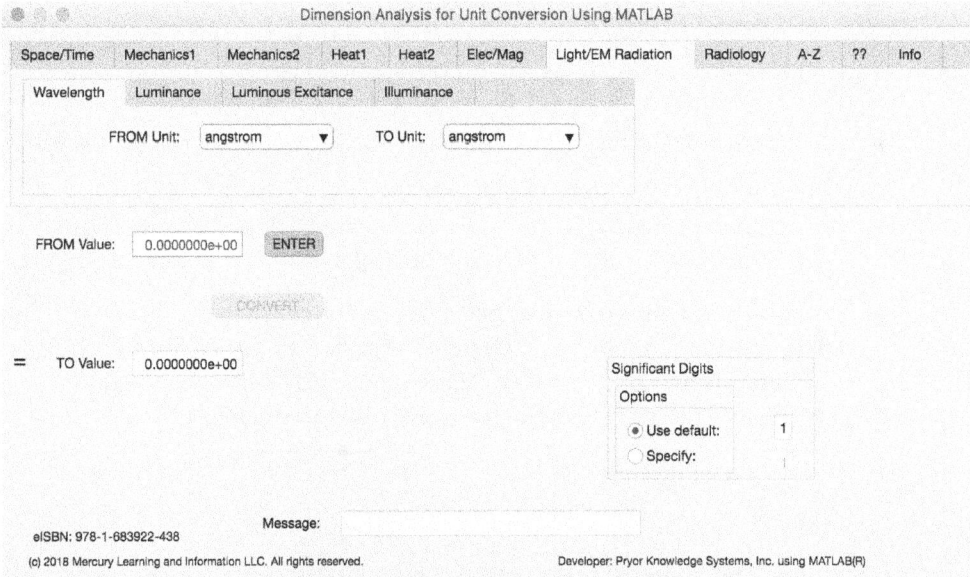

FIGURE 3.7-0 Unit Converter Using MATLAB App - Front Panel Light/EM Radiation Tab.

Wavelength Tab

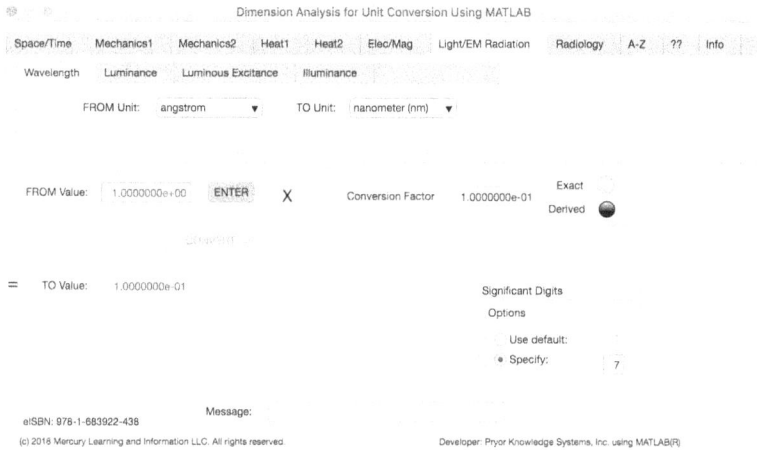

FIGURE 3.7-1 Unit Converter Using MATLAB App - Light/EM Radiation Wavelength Tab.

The first Minor Tab under the Light/EM Radiation Major Tab is the Wavelength Tab (see Figure 3.7-1). This example shows the conversion of angstrom (Å) to nanometer (nm). In this example, the amplitude of angstrom (Å) is set equal to one (1.0). The conversion equation is shown in Example 3.7-1.

angstrom (Å) (=1.0) to nanometer (nm) Example 3.7-1:

$$nm = Å * m/Å * nm/m = Å * 1.0E\text{-}10 * 1.0E9 = Å * 1.0E\text{-}1$$

Where: nm = the number of nanometer calculated
      Å = the number of angstrom to be converted (1.0)
    m/Å = the conversion factor from angstrom to meter (1.0e-10)
   nm/m = the conversion factor from meter to nanometer (1.0E9)

And:

The resulting solution is:

$$nm = 1.0 * 1.0E\text{-}1 = 1.0E\text{-}1 \text{ nanometer} \tag{3.7-1}$$

The Light/EM Radiation Wavelength Tab allows the bidirectional selection of the following wavelength dimensions:

Major-Tab	Minor-Tab	Units on Pull-Down
Light/EM Radiation		
	Wavelength	
		angstrom (Å)
		nanometer (nm)

Luminance Tab

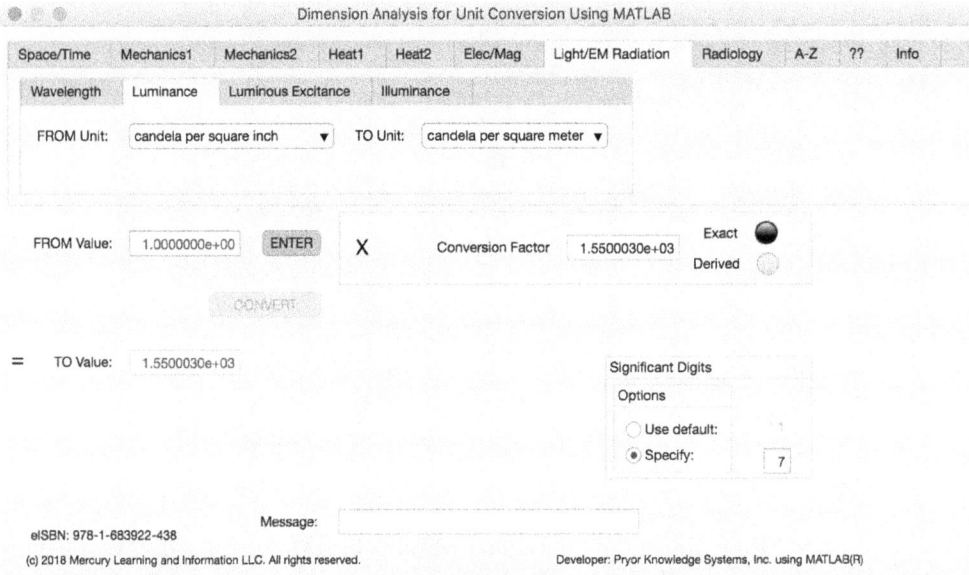

FIGURE 3.7-2 Unit Converter Using MATLAB App - Light/EM Radiation Luminance Tab.

The second Minor Tab under the Light/EM Radiation Major Tab is the Luminance Tab (see Figure 3.7-2). This example shows the conversion of candela per square inch (cd/in^2) to candela per square meter (cd/m^2). In this example, the amplitude of candela per square inch (cd/in^2) is set equal to one (1.0). The conversion equation is shown in Example 3.7-2.

candela per square inch (cd/in^2) (-1.0) to candela per square meter (cd/m^2) Example 3.7-2:

$$cd/m^2 = cd/in^2 * in^2/ m^2$$
$$= cd/in^2 * 1/6.4516E\text{-}4$$
$$= cd/in^2 * 1.550003E\text{+}3$$

Where: cd/m^2 = the number of candela per square meter calculated
cd/in^2 = the number of candela per square inch to be converted (1.0)
in^2/ m^2 = the conversion factor from square meter to square inch
(1.550003E+3)

And:

The resulting solution is:

cd/m^2 = 1.0 * 1.550003E+3 = 1.550003E+3 candela
per square meter   (3.7-2)

The Light/EM Radiation Luminance Tab allows the bidirectional selection of the following luminance dimensions:

Major-Tab	**Minor-Tab**	**Units on Pull-Down**
Light/EM Radiation		
	Luminance	
		lambert (L)
		candela per square inch (cd/in^2)
		footlambert
		candela per square meter (cd/m^2)

Luminous Excitance Tab

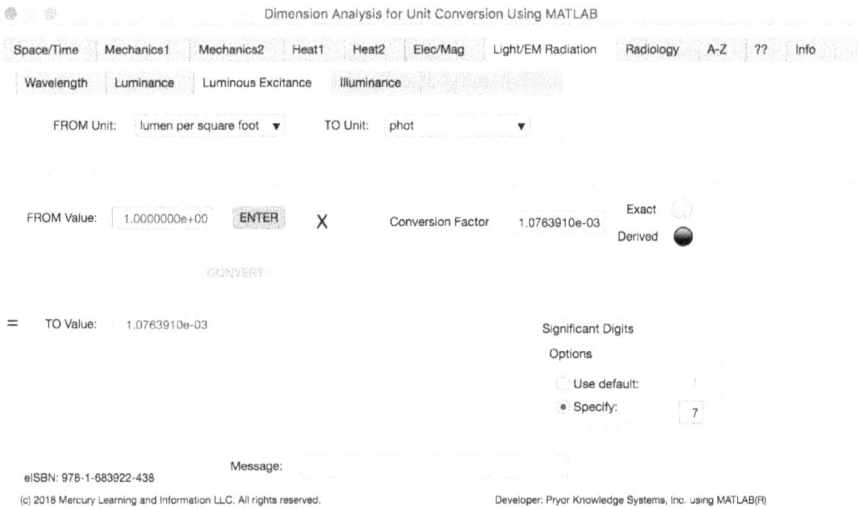

FIGURE 3.7-3 Unit Converter Using MATLAB App - Light/EM Radiation Luminous Excitance Tab.

The third Minor Tab under the Light/EM Radiation Major Tab is the Luminous Excitance Tab (see Figure 3.7-3). This example shows the conversion of lumen per square foot (lm/ft^2) to phot (ph = lumen per square centimeter). In this example, the amplitude of lumen per square foot (lm/ft^2) is set equal to one (1.0). The conversion equation is shown in Example 3.7-3.

lumen per square foot (lm/ft^2) to phot (ph) Example 3.7-3:

$$ph = lm/ft^2 * ft^2/m^2 * m^2/cm^2$$
$$= lm/ft^2 * 1/0.09290304 * 1/10^4$$
$$= lm/ft^2 * 0.00107639 = lm/ft^2 * 1.07639E\text{-}3$$

Where: ph = the number of phot calculated

lm/ft2 = the number of lumen per square foot to be converted (1.0)

in²/m² = the conversion factor from square meter to square foot (1/0.09290304)

m²/cm² = the conversion factor from square centimeter to square meter (1/10⁴)

And:

The resulting solution is:

$$ph = 1.0 * 1.07639E\text{-}3 = 1.07639E\text{-}3 \text{ phot} \tag{3.7-3}$$

The Light/EM Radiation Luminous Excitance Tab allows the bidirectional selection of the following Luminous Excitance dimensions:

Major-Tab	Minor-Tab	Units on Pull-Down
Light/EM Radiation		
	Luminous Excitance	
		lumen per square foot (lm/ft²)
		phot (ph)
		lux (lx)

Illuminance Tab

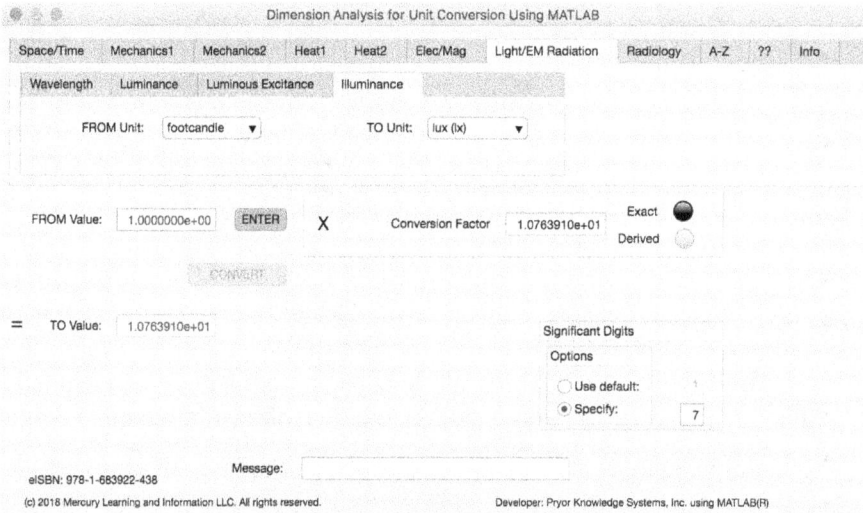

FIGURE 3.7-4 Unit Converter Using MATLAB App - Light/EM Radiation Illuminance Tab.

The fourth Minor Tab under the Light/EM Radiation Major Tab is the Illuminance Tab (see Figure 3.7-4). This example shows the conversion of footcandle (fc = lumen per square foot) to lux (lx = lumen per square meter). In this example, the amplitude of footcandle (fc) is set equal to one (1.0). The conversion equation is shown in Example 3.7-4.

footcandle (fc) to lux (lx) Example 3.7-4:

$$lx = fc \ast ft^2/m^2$$
$$= fc \ast 1/0.09290304$$
$$= fc \ast 10.76391$$
$$= fc \ast 1.076391E+1$$

Where: lx = the number of lux calculated
   fc = the number of footcandle to be converted (1.0)
   ft^2/m^2 = the conversion factor from square meter to square foot
      (1.07639E+1)

And:

The resulting solution is:

$$lx = 1.0 \ast 1.07639E+1 = 1.07639E+1 \; lux \hspace{3cm} (3.7\text{-}4)$$

The Light/EM Radiation Illuminance Tab allows the bidirectional selection of the following illuminance dimensions:

Major-Tab	Minor-Tab	Units on Pull-Down
Light/EM Radiation		
	Illuminance	
		footcandle
		lux (lx)

3.8   QUANTITIES OF RADIOLOGY

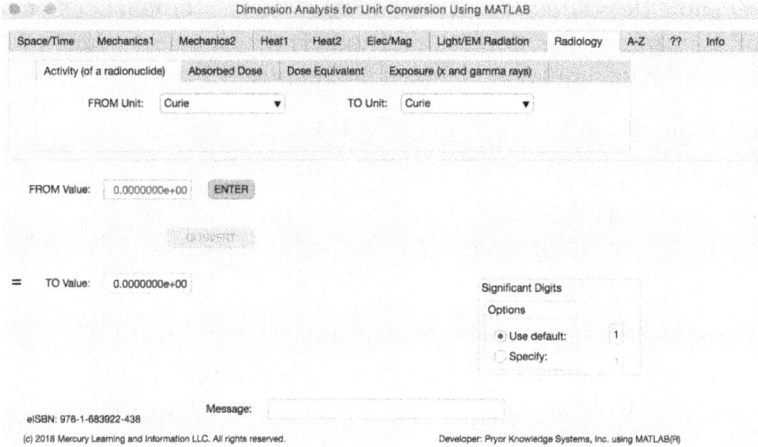

FIGURE 3.8-0 Unit Converter Using MATLAB App - Front Panel Radiology Tab.

Figure 3.8-0 shows the Unit Converter MATLAB App Front Panel with the Radiology Tab selected. Bidirectional matrices were created and installed within this app to allow the user the ability to easily select via pull-down menus all of the desired conversion pairs required. Each of the Tabs available in the App is explored herein.

Activity (of a radionuclide) Tab

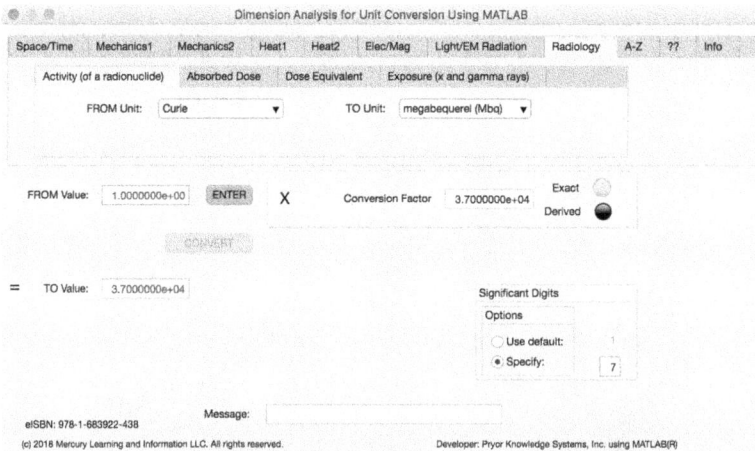

FIGURE 3.8-1 Unit Converter Using MATLAB App - Radiology Activity (of a radionuclide) Tab.

The first Minor Tab under the Radiology Major Tab is the Activity (of a radionuclide) Tab (see Figure 3.8-1). This example shows the conversion of Curie (ci) = 3.7×10^{10} decays per second) to megabequerel (MBq; 1 Bq = 1 decay per second). In this example, the amplitude of Curie (ci) is set equal to one (1.0). The conversion equation is shown in Example 3.8-1.

Curie (ci) to megabequerel (MBq) Example 3.8-1:

$$MBq = ci * Bq/ci * MBq/Bq$$
$$= ci * = ci * 3.7E+10 * 1.0E-6$$
$$= ci * 3.7E+4$$

Where: MBq = the number of megabequerel calculated
    ci = the number of Curie to be converted (1.0)
    Bq/ci = the conversion factor from Curie to bequerel (3.7E+10)
    MBq/Bq = the conversion factor from bequerel to megabequerel (1.0E-6)

And:

The resulting solution is:

$$MBq = 1.0 * 3.7E+4 = 3.7E+4 \text{ megabequerel} \qquad (3.8\text{-}1)$$

The Radiology Activity (of a radionuclide) Tab allows the bidirectional selection of the following Activity (of a radionuclide) dimensions:

Major-Tab	Minor-Tab	Units on Pull-Down
Radiology		
	Activity (of a radionuclide)	
		Curie
		megabequerel (MBq)

Absorbed Dose Tab

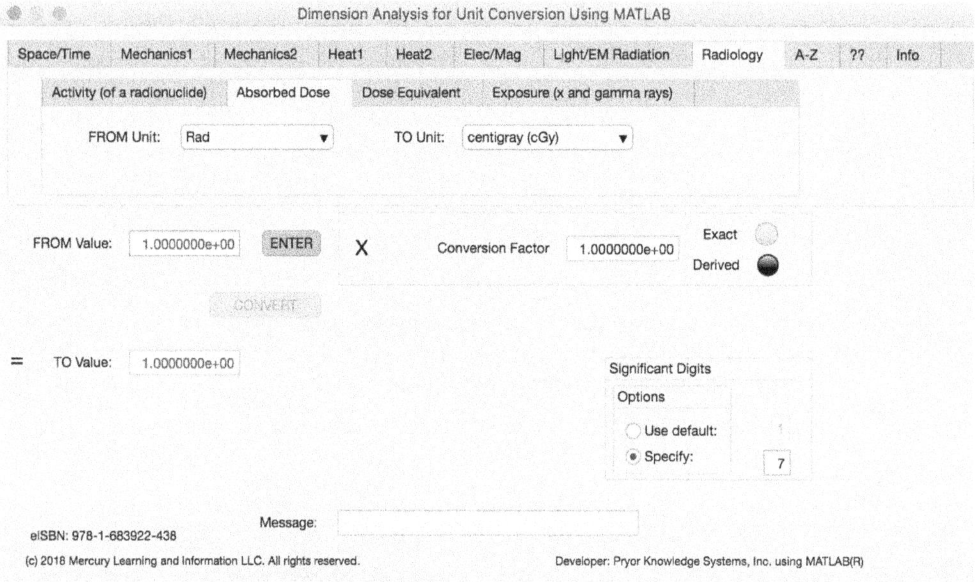

FIGURE 3.8-2 Unit Converter Using MATLAB App - Radiology Absorbed Dose Tab.

The second Minor Tab under the Radiology Major Tab is the Absorbed Dose Tab (see Figure 3.8-2). This example shows the conversion of Rad to centigray (cGy). In this example, the amplitude of Rad is set equal to one (1.0). The conversion equation is shown in Example 3.8-2.

Rad to centigray (cGy) Example 3.8-2:

$$cGy = rad * Gy/rad * cGy/Gy$$
$$= rad * 0.01 * 1.0E+2$$
$$= rad * 1.0E0$$

Where: cGy = the number of centigray calculated
Rad = the number of Rad to be converted (1.0)
Gy/Rad = the conversion factor from Rad to gray (0.01)
cGy/Gy = the conversion factor from gray to centigray (1.0E+2)

And:

The resulting solution is:

$$cGy = 1.0 * 1.0E+0 = 1.0E0 \text{ centigray} \tag{3.8-2}$$

The Radiology Absorbed Dose Tab allows the bidirectional selection of the following absorbed dose dimensions:

Major-Tab	**Minor-Tab**	**Units on Pull-Down**
Radiology		
	Absorbed Dose	
		Rad (rad)
		gray (Gy)
		centigray (cGy)

Dose Equivalent

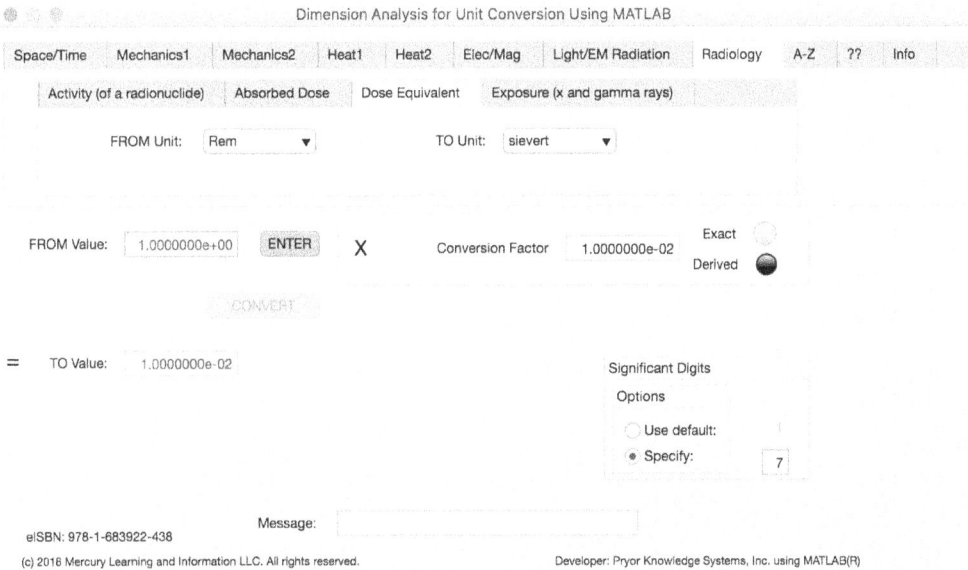

FIGURE 3.8-3 Unit Converter Using MATLAB App - Radiology Dose Equivalent Tab.

The third Minor Tab under the Radiology Major Tab is the Dose Equivalent Tab (see Figure 3.8-3). This example shows the conversion of Rem (rem) to sievert (Sv). In this example, the amplitude of Rem (rem) is set equal to one (1.0). The conversion equation is shown in Example 3.8-3.

Rem (rem) to sievert (Sv) Example 3.8-3:

$$Sv = rem * Sv/rem = rem * 0.01 = rem * 1.0E\text{-}2$$

Where: Sv = the number of sievert calculated
    rem = the number of Rem to be converted (1.0)
    Sv/rem = the conversion factor from Rem to sievert (0.01)

And:

The resulting solution is:

S = 1.0 * 1.0E-2 = 1.0E-2 sievert   (3.8-3)

The Radiology Dose Equivalent Tab allows the bidirectional selection of the following dose equivalent dimensions:

Major-Tab	**Minor-Tab**	**Units on Pull-Down**
Radiology		
	Dose Equivalent	
		Rem (rem)
		sievert (Sv)

Exposure (x and gamma rays) Tab

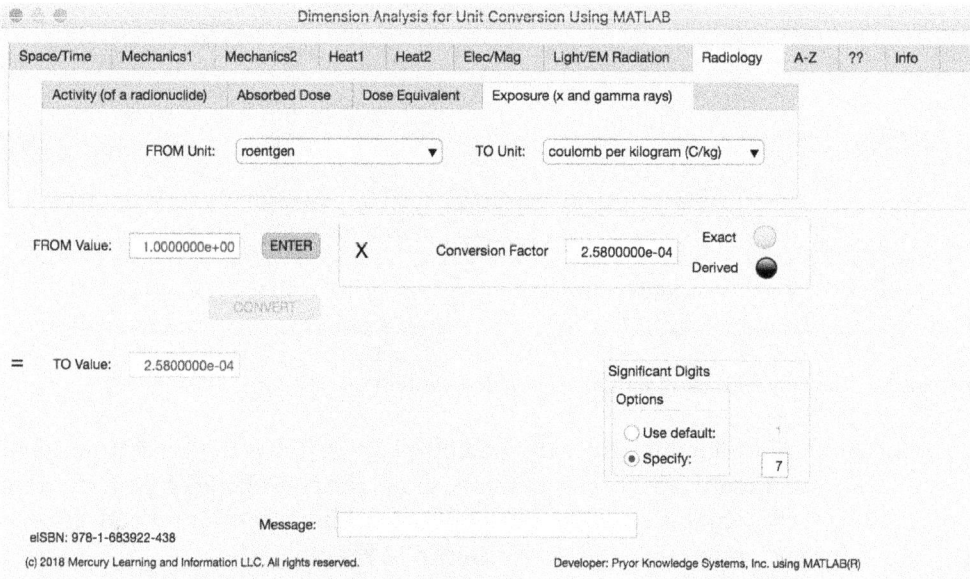

FIGURE 3.8-4 Unit Converter Using MATLAB App - Radiology Exposure (x and gamma rays) Tab.

The fourth Minor Tab under the Radiology Major Tab is the Exposure (x and gamma rays) Tab (see Figure 3.8-4). This example shows the conversion of roentgen (R = lumen per square foot) to coulomb per kilogram (C/kg = lumen per square meter). In this example, the amplitude of roentgen (R) is set equal to one (1.0). The conversion equation is shown in Example 3.8-4.

roentgen (R) to coulomb per kilogram (C/kg) Example 3.8-4:

$$C/kg = R \ ° \ (C/kg)/R = R \ ° \ 0.000258 = R \ ° \ 2.58E\text{-}4$$

Where: C/kg = the number of coulomb per kilogram calculated
       R = the number of roentgen to be converted (1.0)
  (C/kg)/R = the conversion factor from roentgen to coulomb per kilogram (0.000258)

And:

The resulting solution is:

$$C/kg = 1.0 \ ° \ 2.58E\text{-}4 = 2.58E\text{-}4 \text{ coulomb per kilogram} \qquad (3.8\text{-}4)$$

The Radiology Exposure (x and gamma rays) Tab allows the bidirectional selection of the following exposure (x and gamma rays) dimensions:

Major-Tab	**Minor-Tab**	**Units on Pull-Down**
Radiology		
	Exposure (x and gamma rays)	
		roentgen (R)
		coulomb per kilogram (C/kg)

CHAPTER 4

AN INTRODUCTION TO UNACCEPTABLE DIMENSIONAL UNITS

Table 4-1 shows examples of centimeter-gram-second (CGS) units that have special names. Such specially named CGS units are not accepted for general use in commerce with the SI. Also, none of the other units of the various CGS systems of units, which includes the CGS Electrostatic (ESU), CGS Electromagnetic (EMU), and CGS Gaussian systems, are accepted for general use in commerce with the SI except such units as the centimeter, gram, and second that are also defined in the SI.

TABLE 4-1

Name	Symbol	Value in SI Units
erg	erg	10^{-7} J
dyne	dyn	10^{-5} N
poise	P	0.1 Pa°s
stokes	St	10^{-4} m^2/s
gauss	Gs, G	10^{-4} T
oersted	Oe	$(10^3 / 4\pi)$°A/m
maxwell	Mx	10^{-8} Wb
stilb	sb	1 cd/cm^2
phot	ph	10^4 lux
gal	Gal	1 cm/s^2

Table 4-2 shows other unacceptable special name units.

TABLE 4-2

Name	Symbol	Value in SI Units
fermi	fermi	$10^\wedge\text{-}15$ m
photometric carat	metric carat	$2° \; 10^\wedge\text{-}4$ kg
torr	Torr	101325/760 Pa
standard atmosphere	atm	101325 Pa
kilogram-force	kgf	9.80665 N
micron	μ	$10^\wedge\text{-}6$ m
calorie	cal$_{th}$	4.184 J
x unit	xu	$1.002°10^\wedge\text{-}13$ m
stere	st	$1 \; \text{m}^\wedge 3$

AN INTRODUCTION TO THE INTERNATIONAL SYSTEM OF UNITS (SI) – Conversion Factors for General Use (Unidirectional Alphabetical Conversion App Section)

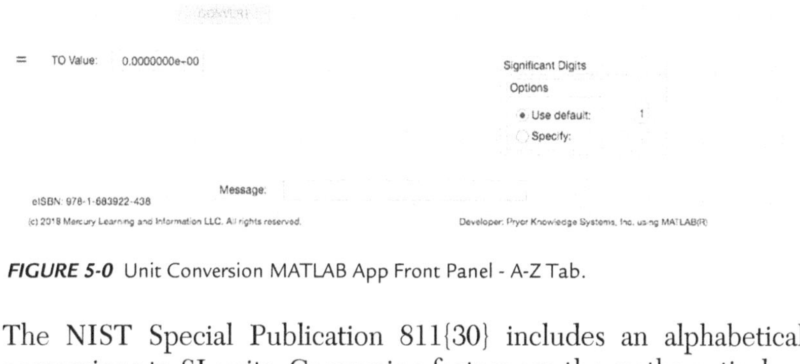

FIGURE 5-0 Unit Conversion MATLAB App Front Panel - A-Z Tab.

The NIST Special Publication 811{30} includes an alphabetical list of conversions to SI units. Conversion factors are the mathematical multiplier or divisor and associated units that allow any compatible measurements to be converted from one system of units to another system of units (e.g., inches to centimeters or vice versa). From-units listed, except those preceded by an asterisk, are not to be used in NIST publications.

Figure 5-0 shows the Unit Converter MATLAB App Front Panel with the A-Z Tab selected. Unidirectional matrices were created and installed within this app to allow the user the ability to easily select via pull-down menus the desired conversion pair. Each of the Tabs available in the App is explored herein.

5.1 A TAB

The first Minor Tab under the A-Z Major Tab is the A Tab (see Figure 5.1-1). This example demonstrates the conversion of ampere hour (A°h) to coulomb (C). In this example, the amplitude of ampere hours is set equal to one (1.0). The conversion equation is shown in Example 5.1-1 {37}.

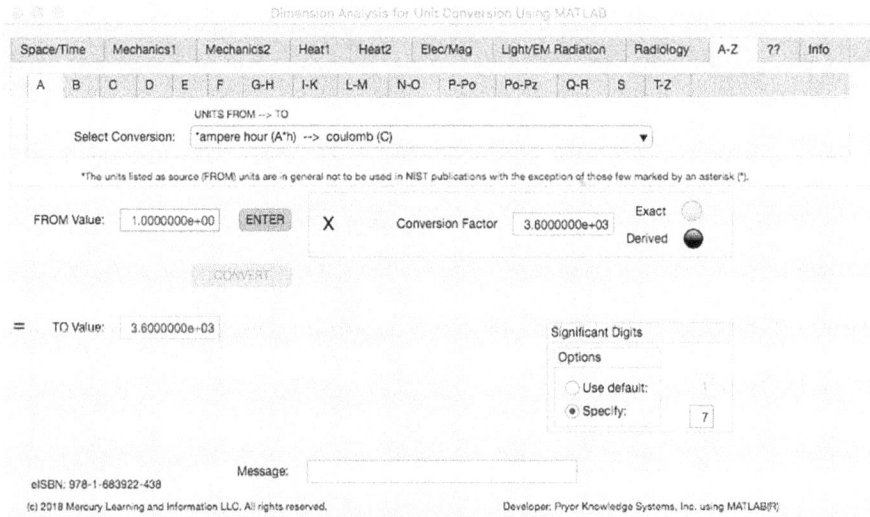

FIGURE 5.1-1 Unit Converter Using MATLAB App A-Z - A Tab.

ampere hour (A°h) (=1.0) to coulomb (C) Example 5.1-1:

$$C = A * h * C/s * 1/A * s/min * min/h$$
$$= A * h * 60 * 60$$
$$= A * h * 3.6E3$$

Where: C = the number of coulombs calculated

A ° h = the number of ampere hours to be converted (1.0)

C/s ° 1/A = the conversion factor from ampere to coulomb per second (1.0)

s/min = the conversion factor from minute to second (**60**)

min/h = the conversion factor from hour to minute (**60**)

And:

The resulting solution is:

C = 1.0 ° 3.6E3 = 3.6E3 coulomb$\qquad$(5.1-1)

The A-Z A Tab allows the unidirectional selection of the following dimensions:

Major/Minor-Tab **From/To Units on Pull-Down**

A-Z / A

abampere to ampere (A)

abcoulomb to coulomb (C)

abfarad to farad (F)

abhenry to henry (H)

abmho to siemens (S)

abohm to ohm (Ω)

abvolt to volt (V)

acceleration in free fall, standard (g)

to meter per second squared (m/s²)

acre (based on U.S. Survey foot) to square meter (m²)

acre foot (based on U.S. survey foot to cubic meter (m³)

°ampere hour (a ° h) to coulomb (C)

angstrom (Å) to meter (m)

angstrom (Å) to nanometer (nm)

are (a) to square meter

°astronomical unit (ua) to meter (m)

atmosphere, standard (atm) to pascal (Pa)

atmosphere, standard (atm) to kilopascal (kPa)

atmosphere, technical (at) to pascal (Pa)

atmosphere, technical (at) to kilopascal (kPa)

5.2   B TAB

The second Minor Tab under the A-Z Major Tab is the B Tab (see Figure 5.2-1). This example shows the conversion of bushel (bu) to cubic meter (m^3). In this example, the amplitude of bushel is set equal to one (1.0). The conversion equation is shown in Example 5.2-1.

bushel (US) (bu) (=1.0) to cubic meter (m^3) Example 5.2-1:

> m^3 = bu * m^3/ bu = bu * 3.523907E-2

Where: m^3 = the number of cubic meters calculated
  bu = the starting number of bushels (1.0)
  m^3/ bu = the conversion factor from bushel to cubic meter
    (3.523907E-2)

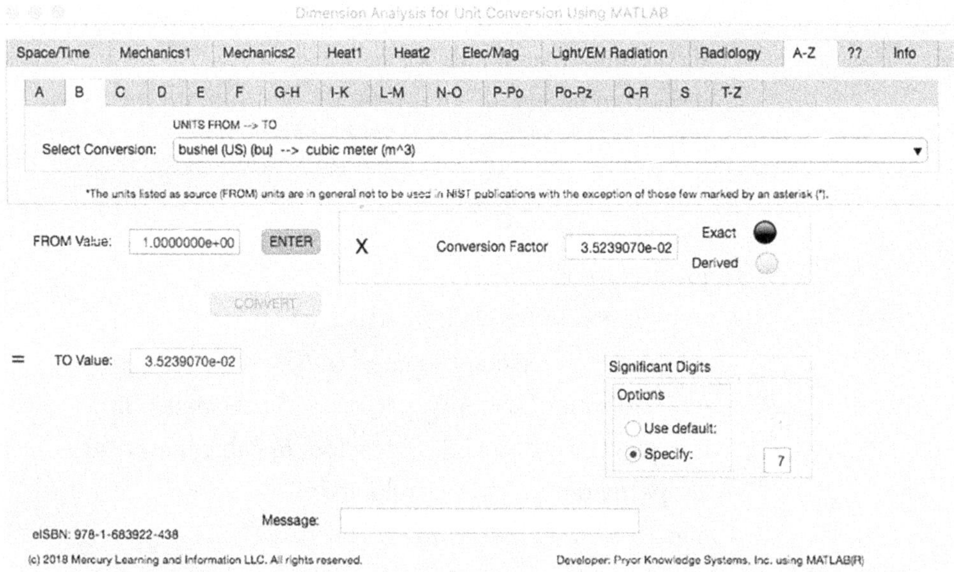

FIGURE 5.2-1 Unit Converter Using MATLAB App A-Z - B Tab.

And:

The resulting solution is:

> m^3 = 1.0 * 3.523907E-2 = 3.523907E-2 cubic meter   (5.2-1)

The A-Z B Tab allows the unidirectional selection of the following dimensions:

Major/Minor-Tab   From/To Units on Pull-Down
A-Z / B

°bar (bar) to pascal (Pa)

°bar (bar) to kilopascal (kPa)

°barn (b) to square meter (m^2)

barrel [for petroleum, 42 gallons (US)] (bbl)
      to cubic meter (m^3)

barrel [for petroleum, 42 gallons (US)] (bbl) to liter (L)

biot (Bi) to ampere (A)

British thermal unit IT (Btu IT) to joule (J)

British thermal unit th (Btu IT) to joule (J)

British thermal unit (mean) (Btu) to joule (J)

British thermal unit (39 degF) to joule (J)

British thermal unit (59 degF) (Btu) to joule (J)

British thermal unit (60 degF) (Btu) to joule (J)

Btu IT foot per hour square foot degree Fahrenheit
                    [Btu IT°ft/(h°ft^2°degF)]
      to watt per meter kelvin [W / (m ° K)]

Btu th foot per hour square foot degree Fahrenheit
                    [Btu IT°ft/(h°ft^2°degF)]
      to watt per meter kelvin [W / (m ° K)]

Btu IT inch per hour square foot degree Fahrenheit
                    [Btu IT°in/(h°ft^2°degF)]
      to watt per meter kelvin [W / (m ° K)]

Btu th inch per hour square foot degree Fahrenheit
                    [Btu th°in/(h°ft^2°degF)]
      to watt per meter kelvin [W / (m ° K)]

Btu IT inch per second square foot degree Fahrenheit
                    [Btu IT°in/(s°ft^2°degF)]
      to watt per meter kelvin [W / (m ° K)]

Major/Minor-Tab   From/To Units on Pull-Down

A-Z / B

Btu th inch per second square foot degree Fahrenheit
[Btu th*in/(s*ft^2*degF)]
to watt per meter kelvin [W / (m * K)]

Btu IT per cubic foot (Btu IT/ft^3)
to joule per cubic meter (J / m^3)

Btu th per cubic foot (Btu th/ft^3)
to joule per cubic meter (J / m^3)

Btu IT per degree Fahrenheit (Btu IT/degF)
to joule per kelvin (J / K)

Btu th per degree Fahrenheit (Btu th/degF)
to joule per kelvin (J/K)

Btu IT per degree Rankine (Btu IT/degR)
to joule per kelvin (J/K)

Btu th per degree Rankine (Btu th/degR)
to joule per kelvin (J/K)

Btu IT per hour (Btu IT/h) to watt (W)

Btu th per hour (Btu th/h) to watt (W)

Btu IT per hour square foot degree Fahrenheit
[Btu IT/(h*ft^2*degF)]
to watt per square meter kelvin [W/(m^2* K)]

Btu th per hour square foot degree Fahrenheit
[Btu th/(h*ft^2*degF)]
to watt per square meter kelvin [W/(m^2* K)]

Btu th per minute (Btu th/min) to watt (W)

Btu IT per pound (Btu IT/lb) to joule per kilogram (J/kg)

Btu th per pound (Btu th/lb) to joule per kilogram (J/kg)

Btu IT per pound degree Fahrenheit [Btu IT/(lb*degF)]
to joule per kilogram kelvin [J/(kg*K)]

Btu th per pound degree Fahrenheit [Btu th/(lb*degF)]
to joule per kilogram kelvin [J/(kg*K)]

Major/Minor-Tab   **From/To Units on Pull-Down**
A-Z / B

Btu IT per pound degree Rankine [Btu IT/(lb°degR)]
   to joule per kilogram kelvin [J/(kg°K)]
Btu th per pound degree Rankine [Btu th/(lb°degR)]
   to joule per kilogram kelvin [J/(kg°K)]
Btu IT per second (Btu IT/s) to watt (W)
Btu th per second (Btu th/s) to watt (W)
Btu IT per second square foot degree Fahrenheit
                        [Btu IT/(s°ft^2°degF)]
   to watt per square meter kelvin [W/(m^2°K)]
Btu th per second square foot degree Fahrenheit
                        [Btu th/(s°ft^2°degF)]
   to watt per square meter kelvin [W/(m^2°K)]
Btu IT per square foot [Btu IT/(ft^2)]
   to joule per square meter [J/(m^2)]
Btu th per square foot [Btu th/(ft^2)]
   to joule per square meter [J/(m^2)]
Btu IT per square foot hour [Btu IT/(ft^2°h)]
   to watt per square meter [W/(m^2)]
Btu th per square foot hour [Btu th/(ft^2°h)]
   to watt per square meter [W/(m^2)]
Btu th per square foot minute [Btu th/(ft^2°min)]
   to watt per square meter [W/(m^2)]
Btu IT per square foot second [Btu IT/(ft^2°s)]
   to watt per square meter [W/(m^2)]
Btu th per square foot second [Btu th/(ft^2°s)]
   to watt per square meter [W/(m^2)]
Btu th per square inch second [Btu th/(in^2°s)]
   to watt per square meter [W/(m^2)]
bushel (US) (bu) to cubic meter (m^3)
bushel (US) (bu) to liter (L)

5.3   C TAB

The third Minor Tab under the A-Z Major Tab is the C Tab (see Figure 5.3-1). This example shows the conversion of cubic yard (yd^3) to cubic meter (m^3). In this example, the amplitude of cubic yard is set equal to one (1.0). The conversion equation is shown in Example 5.3-1.

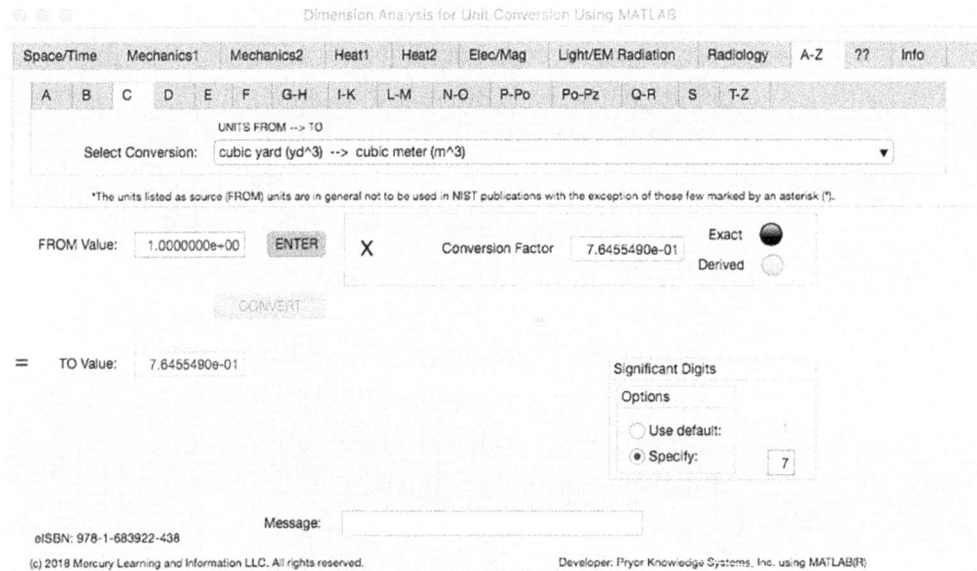

FIGURE 5.3-1 Unit Converter Using MATLAB App A-Z - C Tab.

cubic yard (yd^3) (=1.0) to cubic meter (m^3) Example 5.3-1:

m^3 = yd^3 * m^3/yd^3 = yd^3 * 7.645549E-1

Where: m^3 = the number of cubic meters calculated
yd^3 = the number of cubic yards to be converted (1.0)
m^3/yd^3 = the conversion factor from cubic yard to cubic meter
(7.645549E-1)

And:

The resulting solution is:

m^3 = 1.0 * 7.645549E-1 = 7.645549E-1 cubic meter   (5.3-1)

The A-Z / C Tab allows the unidirectional selection of the following dimensions:

Major/Minor-Tab   **From/To Units on Pull-Down**

A-Z / C

calorie$_{IT}$ (cal$_{IT}$) to joule (J)

calorie$_{th}$ (cal$_{th}$) to joule (J)

calorie (cal) (mean) to joule (J)

calorie (15 °C) (cal) to joule (J)

calorie (20 °C) (cal) to joule (J)

calorie$_{IT}$, kilogram (nutrition) to joule (J)

calorie$_{th}$, kilogram (nutrition) to joule (J)

calorie (mean), kilogram (nutrition) to joule (J)

calorie$_{th}$ per centimeter second degree Celsius [cal$_{th}$/(cm•s•°C)]
   to watt per meter kelvin [W/(m•K)]

calorie$_{IT}$ per gram (cal$_{IT}$/g) to joule per kilogram (J/kg)

calorie$_{th}$ per gram (cal$_{th}$/g) to joule per kilogram (J/kg)

calorie$_{IT}$ per gram degree Celsius [cal$_{IT}$/(g•°C)]
   to joule per kilogram kelvin [J/(kg•K)]

calorie$_{th}$ per gram degree Celsius [cal$_{th}$/(g•°C)]
   to joule per kilogram kelvin [J/(kg•K)]

calorie$_{IT}$ per gram kelvin [cal$_{IT}$/(g•K)]
   to joule per kilogram kelvin [J/(kg•K)]

calorie$_{th}$ per gram kelvin [cal$_{th}$/(g•K)]
   to joule per kilogram kelvin [J/(kg•K)]

calorie$_{th}$ per minute (cal$_{th}$/min) to watt (W)

calorie$_{th}$ per second (cal$_{th}$/s) to watt (W)

calorie$_{th}$ per square centimeter (cal$_{th}$ / cm^2)
   to joule per square meter (J / m^2)

calorie$_{th}$ per square centimeter minute [cal$_{th}$/(cm^2•min)]
   to watt per square meter (watt / m^2)

Major/Minor-Tab   From/To Units on Pull-Down

A-Z / C

calorie$_{th}$ per square centimeter second [cal$_{th}$ / (cm^2 * s)]
  to watt per square meter (watt / m^2)

candela per square inch (cd / in^2)
  to candela per square meter (cd / m^2)

carat, metric to kilogram (kg)

carat, metric to gram (g)

centimeter of mercury (0 degC) to pascal (Pa)

centimeter of mercury (0 degC) to kilopascal (kPa)

centimeter of mercury, conventional (cmHg) to pascal (Pa)

centimeter of mercury, conventional (cmHg)
  to kilopascal (kPa)

centimeter of water (4 degC) to pascal (Pa)

centimeter of water, conventional (cmH2O) to pascal (Pa)

centipoise (cP) to pascal second (Pa * s)

centistokes (cSt) to meter squared per second (m^2 / s)

chain (based on US survey foot) (ch) to meter (m)

circular mil to square meter (m^2)

circular mil to square millimeter (mm^2)

clo to square meter kelvin per watt (m^2 * K / W)

cord (128 ft^3) to cubic meter (m^3)

cubic foot (ft^3) to cubic meter (m^3)

cubic foot per minute (ft^3 / min)
  to cubic meter per second (m^3 / s)

cubic foot per minute (ft^3 / min) to liter per second (L / s)

cubic foot per second (ft^3 / s)
  to cubic meter per second (m^3/s)

cubic inch (in^3) to cubic meter (m^3)

Major/Minor-Tab	**From/To Units on Pull-Down**
A-Z / C	

cubic inch per minute (in^3/min)
  to cubic meter per second (m^3/s)
cubic mile (mi^3) to cubic meter (m^3)
cubic yard (yd^3) to cubic meter (m^3)
cubic yard per minute (yd^3/min)
  to cubic meter per second (m^3/s)
cup (US) to cubic meter (m^3)
cup (US) to liter (L)
cup (US) to milliliter (mL)
*Curie (Ci) to becquerel (Bq)

5.4   D TAB

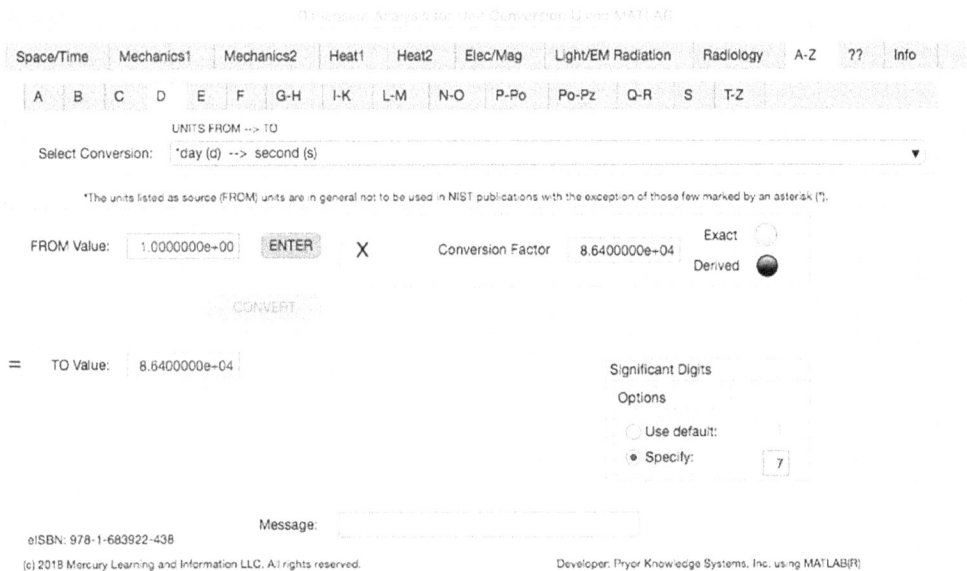

FIGURE 5.4-1 Unit Converter Using MATLAB App A-Z - D Tab.

The fourth Minor Tab under the A-Z Major Tab is the D Tab (see Figure 5.4-1). This example shows the conversion of day (d) to second (s). In this example, the amplitude of day is set equal to one (1.0). The conversion equation is shown in Example 5.4-4.

day (d) (=1.0) to seconds (s) Example 5.4-4:

$$s = d * h/d * min/h * s/min$$
$$= d * 24 * 60 * 60$$
$$= d * 8.64E4$$

Where: s = the number of second calculated
      d = the number of day to be converted (1.0)
    h/d = the conversion factor from day to hour (24)
  min/h = the conversion factor from hour to minute (60)
  s/min = the conversion factor from minute to second (60)

And:

The resulting solution is:

$$s = 1.0 * 8.64E4 = 8.64E4 \text{ seconds} \tag{5.4-1}$$

The A-Z D Tab allows the bidirectional selection of the following dimensions:

Major/Minor-Tab From/To Units on Pull-Down
A-Z / D

        darcy to meter squared (m^2)
        *day (d) to second (s)
        day (sidereal) to second (s)
        debye (D) to coulomb meter (C * m)
        *degree (angle) to radian (rad)
        *degree Celsius (temperature interval) (degC) to kelvin (K)
        degree centigrade (temperature interval)
           to degree Celsius (degC)
        degree Fahrenheit (temperature interval) (degF)
           to degree Celsius (degC)
        degree Fahrenheit (temperature interval) (degF)
           to kelvin (K)

Major/Minor-Tab   **From/To Units on Pull-Down**

A-Z / D

degree Fahrenheit hour per British thermal unit IT
(degF ° h/Btu IT)
   to kelvin per watt (K/W)

degree Fahrenheit hour per British thermal unit th
(degF ° h/Btu th)
   to kelvin per watt (K / W)

degree Fahrenheit hour square foot per Btu IT
(degF°h°ft^2/Btu IT)
   to square meter kelvin per watt (m^2°K/W)

degree Fahrenheit hour square foot per Btu th
(degF°h°ft^2/Btu th)
   to square meter kelvin per watt (m^2°K/W)

degree Fahrenheit hour square foot per Btu IT inch
(degF°h°ft^2/Btu IT°in)
   to meter kelvin per watt (m°K/W)

degree Fahrenheit hour square foot per Btu th inch
(degF°h°ft^2/Btu th°in)
   to meter kelvin per watt (m°K/W)

degree Fahrenheit second per Btu IT (degF°s/Btu IT)
   to kelvin per watt (K/W)

degree Fahrenheit second per Btu th (degF°s/Btu th)
   to kelvin per watt (K/W)

degree Rankine (temperature interval) (degR) to kelvin (K)

denier to kilogram per meter (kg/m)

denier to gram per meter (g/m)

dyne (dyn) to newton (N)

dyne centimeter (dyn-cm) to newton meter (N ° m)

dyne per square centimeter (dyn/cm^2) to pascal (Pa)

5.5 E TAB

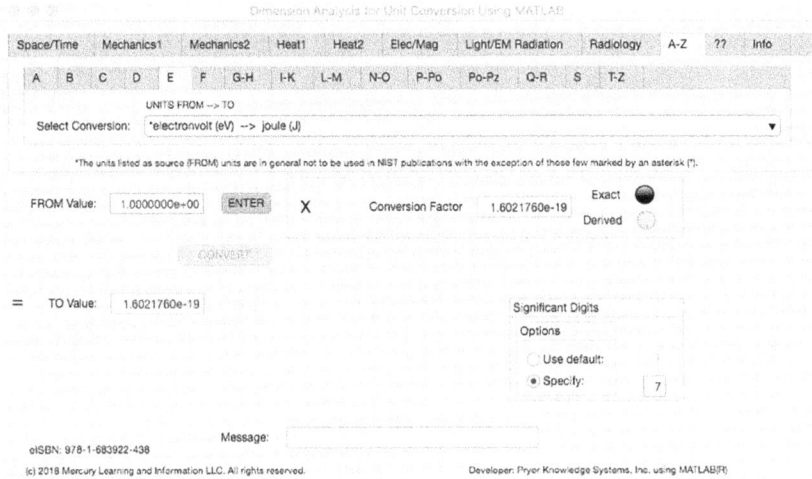

FIGURE 5.5-1 Unit Converter Using MATLAB App A-Z - E Tab.

The fifth Minor Tab under the A-Z Major Tab is the E Tab (see Figure 5.5-1). This example shows the conversion of electron volt (eV) to joule (J). In this example, the amplitude of electron volt (eV) is set equal to one (1.0). The conversion equation is shown in Example 5.5-1.

electronvolt (eV) (=1.0) to joule (J) Example 5.5-1:

$$J = eV * J/eV$$
$$= eV * 1.602176E-19$$

Where: J = the number of joule calculated
eV = the number of electronvolt to be converted (1.0)
J/eV = the conversion factor from electronvolt to joule (1.602176E-19)

And:

The resulting solution is:

$$J = 1.0 * 1.602176E-19 = 1.602176E-19 \text{ joules} \tag{5.5-1}$$

The A-Z E Tab allows the bidirectional selection of the following dimensions:

Major/Minor-Tab   From/To Units on Pull-Down
A-Z / E

*electronvolt (eV) to joule (J)

EMU of capacitance (abfarad) to farad (F)

Major/Minor-Tab	From/To Units on Pull-Down
A-Z / E	

EMU of current (abampere) to ampere (A)

EMU of Electric potential (abvolt) to volt (V)

EMU of inductance (abhenry) to henry (H)

EMU of resistance (abohm) to ohm

erg (erg) to joule (J)

erg per second (erg / s) to watt (W)

erg per square centimeter second [erg / (cm^2 * s)]
  to watt per square meter (W / m^2)

ESU of capacitance (statfarad) to farad (F)

ESU of current (statampere) to ampere (A)

ESU of Electric potential (statvolt) to volt (V)

ESU of inductance (stathenry) to henry (H)

ESU of resistance (statohm) to ohm

5.6 F TAB

The sixth Minor Tab under the A-Z Major Tab is the F Tab (see Figure 5.6-1). This example shows the conversion of foot per minute (ft/min) to meter per second (m/s). In this example, the amplitude of foot per minute is set equal to one (1.0). The conversion equation is shown in Example 5.6-1.

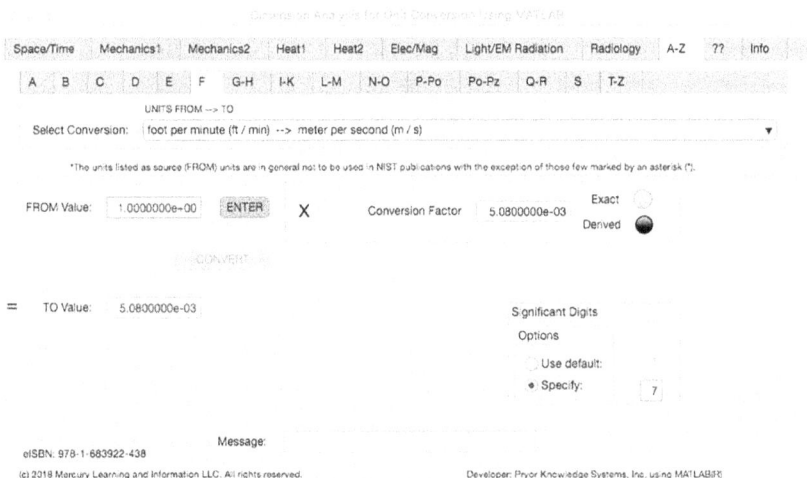

FIGURE 5.6-1 Unit Converter Using MATLAB App A-Z - F Tab.

foot per minute (ft/min) (=1.0) to meter per second (m/s) Example 5.6-1:

m/s = ft/min * m/ft * min/s
   = ft/min * 0.3048 * 1/60
   = ft/min * 5.08E-3

Where: m/s = the number of meter per second calculated
    ft/min = the number of foot per minute to be converted (1.0)
    m/ft = the conversion factor from foot to meter (**0.3048**)
    min/s = the conversion factor from second to minute (**1/60**)

And:

The resulting solution is:

m/s = 1.0 * 0.3048 * 1/60 = 5.08E-3 meter per second   (5.6-1)

The A-Z F Tab allows the unidirectional selection of the following dimensions:

Major/Minor-Tab   From/To Units on Pull-Down
A-Z / F

faraday (based on carbon 12) to coulomb (C)

fathom (based on US survey foot) to meter (m))

fermi to meter (m)

fermi to femtometer (fm)

fluid ounce (US) (fl oz) to cubic meter (m^3)

fluid ounce (US) (fl oz) to milliliter (mL)

foot (ft) to meter (m)

foot (US Survey) (ft) to meter (m)

footcandle to lux (lx)

footlambert to candela per square meter (cd / m^2)

foot of mercury, conventional (ftHg) to pascal (Pa)

foot of mercury, conventional (ftHg) to kilopascal (kPa)

foot of water (39.2 degF) to pascal (Pa)

foot of water (39.2 degF) to kilopascal (kPa)

foot of water, conventional (ftH2O) to pascal (Pa)

foot of water, conventional (ftH2O) to kilopascal (kPa)

foot per hour (ft / h) to meter per second (m / s)

foot per minute (ft / min) to meter per second (m / s)

Major/Minor-Tab	From/To Units on Pull-Down
A-Z / F	

foot per second (ft / s) to meter per second (m / s)

foot per second squared (ft / s^2)
   to meter per second squared (m / s^2)

foot poundal to joule (J)

foot pound-force (ft * lbf) to joule (J)

foot pound-force per hour (ft * lbf / h) to watt (W)

foot pound-force per minute (ft * lbf / min) to watt (W)

foot pound-force per second (ft * lbf / s) to watt (W)

foot to the fourth power (ft^4)
   to meter to the fourth power (m^4)

franklin (Fr) to coulomb (C)

5.7   G-H TAB

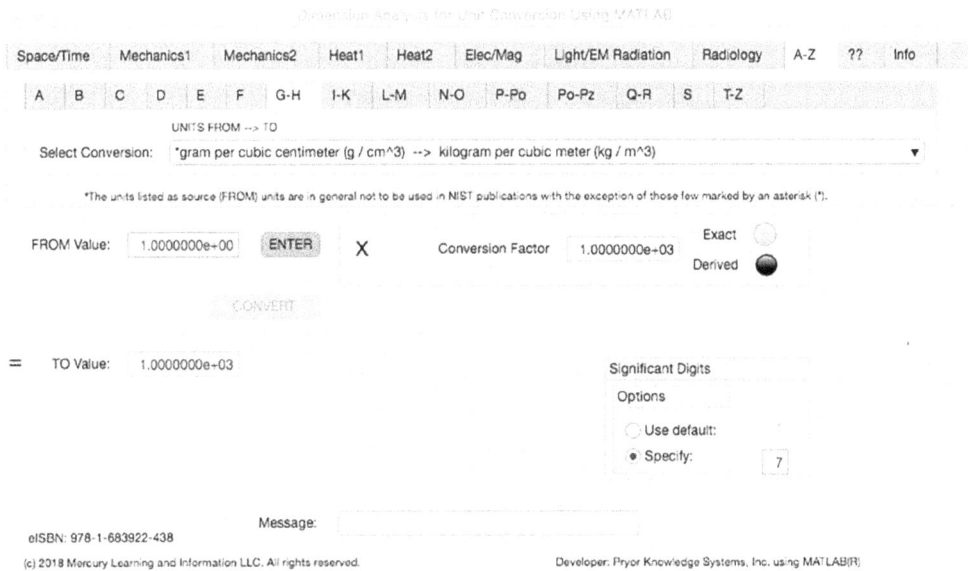

FIGURE 5.7-1 Unit Converter Using MATLAB App A-Z - G-H Tab.

The seventh Minor Tab under the A-Z Major Tab is the G-H Tab (see Figure 5.7-1). This example shows the conversion of gram per cubic centimeter (g/cm³) to kilogram per cubic meter (kg/m³). In this example, the amplitude of gram per cubic centimeter is set equal to one (1.0). The conversion equation is shown in Example 5.7-1.

gram per cubic centimeter (g/cm^3) (=1.0) to kilogram per cubic meter (kg/m^3) Example 5.7-1:

$$kg/m^3 = g/cm^3 \text{ * } kg/g \text{ * } cm^3/m^3$$
$$= g/cm^3 \text{ * } 1.0E\text{-}3 \text{ * } 100^3/1.0^3$$
$$= g/cm^3 \text{ * } 1.0E\text{-}3 \text{ * } 1.0E6$$
$$= g/cm^3 \text{ * } 1.0E3$$

Where: kg/m^3 = the number of kilogram per cubic meter calculated
g/cm^3 = the number of grams per cubic centimeter to be converted (1.0)
kg/g = the conversion factor from gram to kilogram (1.0E-3)
cm^3/m^3 = the conversion factor from cubic meter to cubic centimeter (100^3/1.0^3)

And:

The resulting solution is:

$$kg/m^3 = g/cm^3 \text{ * } 1.0E3 = 1.0 \text{ * } 1.0E3$$
$$= 1.0E3 \text{ kilogram per cubic meter} \qquad (5.7\text{-}1)$$

The A-Z G-H Tab allows the unidirectional selection of the following dimensions:

Major/Minor-Tab	From/To Units on Pull-Down
A-Z / G-H	
	gal (Gal) to meter per second squared (m / s^2)
	gallon [Canadian and U.K. (Imperial)] (gal) to cubic meter (m^3)
	gallon [Canadian and U.K. (Imperial)] (gal) to liter (L)
	gallon (US) (gal) to cubic meter (m^3)
	gallon (US) (gal) to liter (L)
	gallon (US) per day (gal / d) to cubic meter per second (m^3 / s)

Major/Minor-Tab   **From/To Units on Pull-Down**

A-Z / G-H

gallon (US) per day (gal / d) to liter per second (L / s)

gallon (US) per horsepower hour [gal/(hp * h)]
    to cubic meter per joule (m^3/J)

gallon (US) per horsepower hour [gal/(hp * h)]
    to liter per joule (L/J)

gallon (US) per minute (gpm) (gal/min)
    to cubic meter per second (m^3/s)

gallon (US) per minute (gpm) (gal/min)
    to liter per second (L/s)

gamma to tesla (T)

gauss (Gs, G) to tesla (T)

gilbert (Gi) to ampere (A)

gill [Canadian and U.K. (Imperial)] (gi)
    to cubic meter (m^3)

gill [Canadian and U.K. (Imperial)] (gi) to liter (L)

gill (US) (gi) to cubic meter (m^3)

gill (US) (gi) to liter (L)

gon (also called grade) (gon) to radian (rad)

gon (also called grade) (gon) to degree (angle)

grain (gr) to kilogram (kg)

grain (gr) to milligram (mg)

grain per gallon (US) (gr/gal)
    to kilogram per cubic meter (kg/m^3)

grain per gallon (US) (gr/gal)
    to milligram per liter (mg/L)

gram-force per square centimeter (gf/cm^2) to pascal (Pa)

*gram per cubic centimeter (g/cm^3)
    to kilogram per cubic meter (kg/m^3)

*hectare (ha) to square meter (m^2)

Major/Minor-Tab	From/To Units on Pull-Down
A-Z / G-H	
	horsepower (550 ft * lbf / s) (hp) to watt (W)
	horsepower (boiler) to watt (W)
	horsepower (electric) to watt (W)
	horsepower (metric) to watt (W)
	horsepower (U.K.) to watt (W)
	horsepower (water) to watt (W)
	*hour to second (s)
	hour (sidereal) to second (s)
	hundredweight (long, 112 lb) to kilogram (kg)
	hundredweight (short, 100 lb) to kilogram (kg)

5.8   I-K TAB

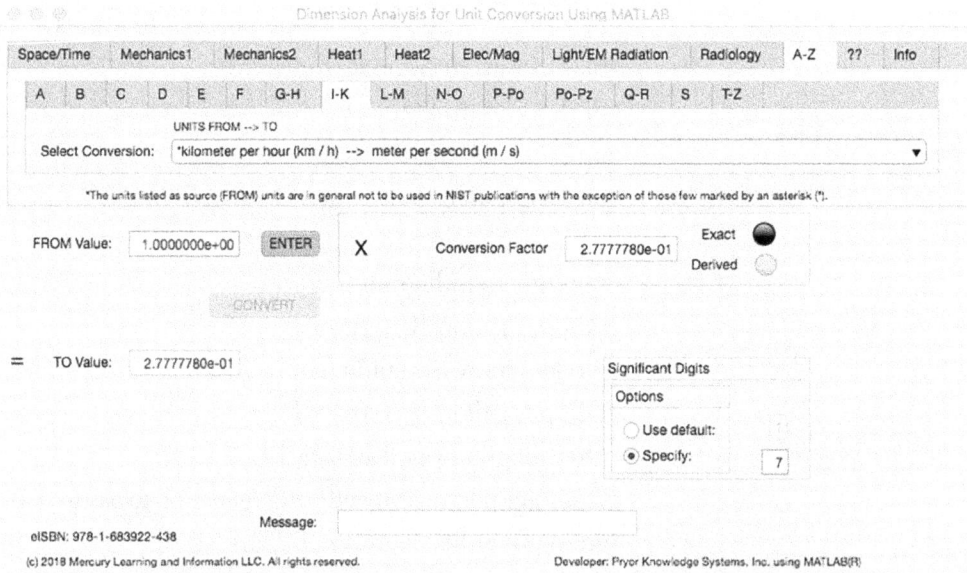

FIGURE 5.8-1 Unit Converter Using MATLAB App A-Z - I-K Tab.

The eighth Minor Tab under the A-Z Major Tab is the I-K Tab (see Figure 5.8-1). This example shows the conversion of kilometer per hour (km/h) to meter per second (m/s). In this example, the amplitude of kilometer per hour (km/h) is set equal to one (1.0). The conversion equation is shown in Example 5.8-1.

kilometer per hour (km/h) (=1.0) to meter per second (m/s) Example 5.8-1:

$$m/s = km/h * h/min * min/s * m/km$$
$$= km/h * 1/60 * 1/60 * 1.0E3$$
$$= km/h * 2.777778E\text{-}1$$

Where: m/s = the number of meter per second calculated
   km/h = the number of kilometer per hour to be converted (1.0)
   h/min = the conversion factor from minutes to hour (1/60)
   min/s = the conversion factor from seconds to minutes (1/60)
   m/km = the conversion factor from kilometers to meters (1.0E3)

And:

The resulting solution is:

$$m/s = km/h * 2.777778E\text{-}1$$
$$= 1.0 * 2.777778E\text{-}1 \text{ kilometers per hour} \tag{5.8-1}$$

The A-Z I-K Tab allows the bidirectional selection of the following dimensions:

Major/Minor-Tab   **From/To Units on Pull-Down**

A-Z / I-K

      inch (in) to meter (m)
      inch (in) to centimeter (cm)
      inch of mercury (32 degF) to pascal (Pa)
      inch of mercury (32 degF) to kilopascal (kPa)
      inch of mercury (60 degF) to pascal (Pa)
      inch of mercury (60 degF) to kilopascal (kPa)
      inch of mercury, conventional (inHg) to pascal (Pa)
      inch of mercury, conventional (inHg) to kilopascal (kPa)
      inch of water (39.2 degF) to pascal (Pa)

Major/Minor-Tab   From/To Units on Pull-Down
A-Z / I-K

inch of water (60 degF) to pascal (Pa)

inch of water, conventional (inH2O) to pascal (Pa)

inch per second (in/s) to meter per second (m/s)

inch per second squared (in/s^2)
   to meter per second squared (m/s^2)

inch to the fourth power (in^4)
   to meter to the fourth power (m^4)

kayser (K) to reciprocal meter (m^-1)

kilocalorie IT (kcal IT) to joule (J)

kilocalorie th (kcal th) to joule (J)

kilocalorie (mean) (kcal) to joule (J)

kilocalorie th per minute (kcal th/min) to watt (W)

kilocalorie th per second (kcal th/s) to watt (W)

kilogram-force (kgf) to newton (N)

kilogram-force meter (kgf * m) to newton meter (N * m)

kilogram-force per square centimeter (kgf/cm^2)
   to pascal (Pa)

kilogram-force per square centimeter (kgf/cm^2)
   to kilopascal (kPa)

kilogram-force per square meter (kgf/m^2) to pascal (Pa)

kilogram-force per square millimeter (kgf/mm^2)
   to pascal (Pa)

kilogram-force per square millimeter (kgf/mm^2)
   to megapascal (Pa)

kilogram-force second squared per meter (kgf * s^2/m)
   to kilogram (kg)

*kilometer per hour (km/h) to meter per second (m/s)

kilopond (kilogram-force) (kp) to newton (N)

Major/Minor-Tab	From/To Units on Pull-Down
A-Z / I-K	

*kilowatt hour (kW * h) to joule (J)

*kilowatt hour (kW * h) to megajoule (MJ)

kip (1 kip = 1 000 lbf) to newton (N)

kip (1 kip = 1 000 lbf) to kilonewton (kN)

kip per square inch (ksi) (kip/in^2) to pascal (Pa)

kip per square inch (ksi) (kip/in^2) to kilopascal (kPa)

*knot (nautical mile per hour) to meter per second (m/s)

5.9   L-M TAB

The ninth Minor Tab under the A-Z Major Tab is the L-M Tab (see Figure 5.9-1). This example demonstrates the conversion of mile (mi) to meter (m). In this example, the amplitude of mile is set equal to one (1.0). The conversion equation is shown in Example 5.9-1.

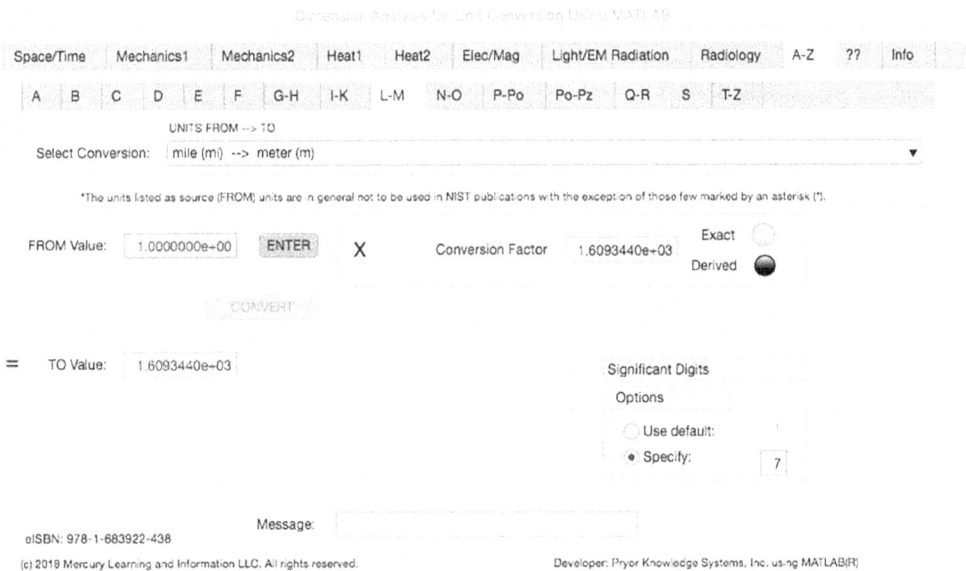

FIGURE 5.9-1 Unit Converter Using MATLAB App A-Z - L-M Tab.

mile (mi) (=1.0) to meter (m) Example 5.9-1:

$$m = mi * km/mi * m/km$$
$$= mi * 1.609344 * 1000$$
$$= mi * 1.609344E3$$

Where: m = the number of meters calculated
      mi = the number of miles to be converted (1.0)
   km/mi = the conversion factor from mile to kilometer (1.609344)
   m/km = the conversion factor from kilometer to meter (1000)

And:

    The resulting solution is:

$$m = 1.0 * 1.609344E3 = 1.609344E3 \text{ meter} \tag{5.9-1}$$

The A-Z L-M Tab allows the unidirectional selection of the following dimensions:

Major/Minor-Tab   **From/To Units on Pull-Down**
A-Z / L-M

          lambert to candela per square meter (cd/m^2)
          langley (cal th/cm^2) to joule per square meter (J/m^2)
          light year (l.y.) to meter (m)
          *liter (L) to cubic meter (m^3)
          lumen per square foot (lm/ft^2) to lux (lx)
          maxwell (Mx) to weber (Wb)
          mho to siemens (S)
          microinch to meter (m)
          microinch to micrometer (um)
          micron (u) to meter (m)
          micron (u) to micrometer (um)
          mil (0.001 in) to meter (m)
          mil (0.001 in) to millimeter (mm)
          mil (angle) to radian (rad)

Major/Minor-Tab	From/To Units on Pull-Down
A-Z / L-M	

mil (angle) to degree, mile (mi) to meter (m)

mile (mi) to kilometer (km)

mile (based on US survey foot) (mi) to meter (m)

mile (based on US survey foot) (mi) to kilometer (km)

*mile, nautical to meter (m)

mile per gallon (US) (mpg) (mi/gal)
   to meter per cubic meter (m/m^3)

mile per gallon (US) (mpg) (mi/gal)
   to kilometer per liter (km / L)

mile per hour (mi/h) to meter per second (m/s)

mile per hour (mi/h) to kilometer per hour (km/h)

mile per minute (mi/min) to meter per second (m/s)

mile per second (mi/s) to meter per second (m/s)

millibar (mbar) to pascal (Pa)

millibar (mbar) to kilopascal (kPa)

*millimeter of mercury, conventional (mmHg)
   to pascal (Pa)

millimeter of water, conventional (mmH2O) to pascal (Pa)

*minute (angle) (') to radian (rad)

minute (min) to second (s)

minute (sidereal) to second (s)

5.10   N-O TAB

The tenth Minor Tab under the A-Z Major Tab is the N-O Tab (see Figure 5.10-1). This example demonstrates the conversion of ounce (avoirdupois) per cubic inch (oz/in^3) to kilogram per cubic meter (kg/m^3). In this example, the amplitude of ounce per cubic inch is set equal to one (1.0). The conversion equation is shown in Example 5.10-1.

FIGURE 5.10-1 Unit Converter Using MATLAB App A-Z - N-O Tab.

ounce (avoirdupois) per cubic inch (oz/in^3) to kilogram per cubic meter (kg/m^3) Example 5.10-1:

kg/m^3 = oz/in^3 * in^3/cm^3 * cm^3/m^3 * g/oz * kg/g
       = oz/in^3 *6.1023744E-2 * 1.0E6 * 2.834952E1 * 1.0E-3
       = oz/in^3 * 1.7299939E3

Where: kg/m^3 = the number of kilogram per cubic meter calculated
       oz/in^3 = the number of ounces per cubic inch to be converted (1.0)
    in^3/cm^3 = the conversion factor from cubic centimeter to cubic inch (6.1023744E-2)
    cm^3/m^3 = the conversion factor from cubic meter to cubic centimeter (1.0E6)
         g/oz = the conversion factor from ounce to gram (2.834952E1)
         kg/g = the conversion factor from gram to kilogram (1.0e-3)

And:

The resulting solution is:

kg/m^3 = 1.0 * 1.7299939E3
       = 1.7299939E3 kilogram per cubic meter   (5.10-1)

The A-Z N-O Tab allows the unidirectional selection of the following dimensions:

Major/Minor-Tab   From/To Units on Pull-Down

A-Z / N-O

oersted (Oe) to ampere per meter (A/m)

*ohm centimeter (ohm * cm) to ohm meter (ohm * m)

ohm circular-mil per foot to ohm meter (ohm * m)

ohm circular-mil per foot
    to ohm square millimeter per meter (ohm * mm^2/m)

ounce (avoirdupois) (oz) to kilogram (kg)

ounce (avoirdupois) (oz) to gram (g)

ounce (troy or apothecary) (oz) to kilogram (kg)

ounce (troy or apothecary) (oz) to gram (g)

ounce [Canadian and U.K. fluid (Imperial)] (fl oz)
    to cubic meter (m^3)

ounce [Canadian and U.K. fluid (Imperial)] (fl oz)
    to milliliter (mL)

ounce (U.S. fluid) (fl oz) to cubic meter (m^3)

ounce (U.S. fluid) (fl oz) to milliliter (mL)

ounce (avoirdupois)-force (ozf) to newton (N)

ounce (avoirdupois)-force inch (ozf * in)
    to newton meter (N * m)

ounce (avoirdupois)-force inch (ozf * in)
    to millinewton meter (mN * m)

ounce (avoirdupois) per cubic inch (oz / in^3)
    to kilogram per cubic meter (kg/m^3)

ounce (avoirdupois) per gallon
                [Canadian and U.K. (Imperial)] (oz/gal)
    to kilogram per cubic meter (kg/m^3)

ounce (avoirdupois) per gallon
                [Canadian and U.K. (Imperial)] (oz/gal)
    to gram per liter (g/L)

ounce (avoirdupois) per gallon (U.S.) (oz/gal)
    to kilogram per cubic meter (kg/m^3)

Major/Minor-Tab From/To Units on Pull-Down

A-Z / N-O

ounce (avoirdupois) per gallon (U.S.) (oz/gal)
to gram per liter (g/L)

ounce (avoirdupois) per square foot (oz/ft^2)
to kilogram per square meter (kg/m^2)

ounce (avoirdupois) per square inch (oz/in^2)
to kilogram per square meter (kg/m^2)

ounce (avoirdupois) per square yard (oz/yd^2)
to kilogram per square meter (kg/m^2)

5.11 P-PO TAB

The eleventh Minor Tab under the A-Z Major Tab is the P-Po Tab (see Figure 5.11-1). This example demonstrates the conversion of pica (computer) (1/6 in) to meter (m). In this example, the amplitude of pica (computer) (1/6 in) is set equal to one (1.0). The conversion equation is shown in Example 5.11-1.

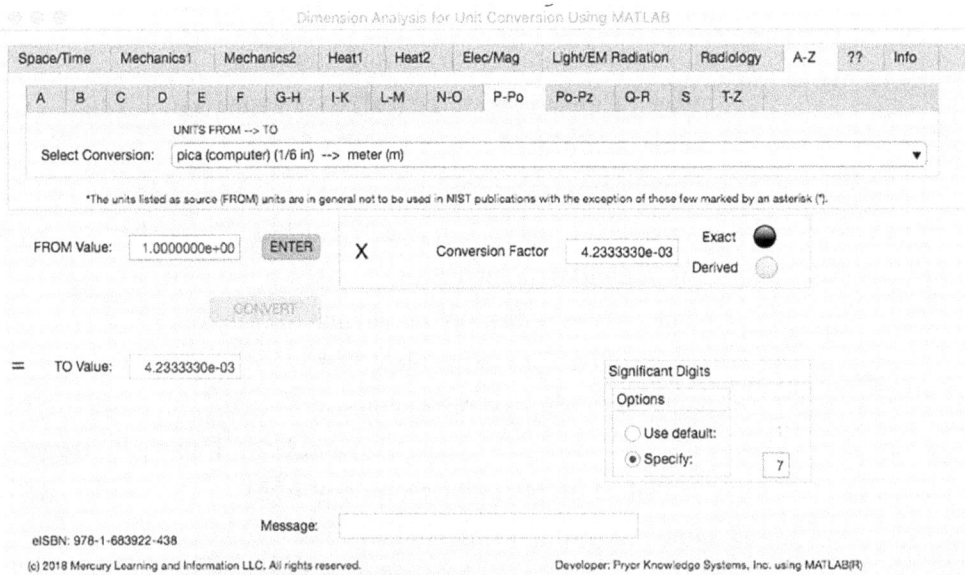

FIGURE 5.11-1 Unit Converter Using MATLAB App A-Z - P-Po Tab.

pica (computer) (1/6 in) to meter (m) Example 5.11-1:

m = pica (computer) ° in/pica (computer) ° cm/in ° m/cm
    = pica (computer) ° 1/6 ° 2.54 ° 1.0E-2
    = pica (computer) ° 4.2333333E-3

Where: m = the number of meter calculated
    pica (computer) = the number of pica to be converted (1.0)
in/pica (computer) = the conversion factor from pica to inch (1/6)
            cm/in = the conversion factor from inch to centimeter (2.54)
            m/cm = the conversion factor from centimeter to meter (1.0E-2)

And:

    The resulting solution is:

m = 1.0 ° 4.2333333E-3 = 4.2333333E-3 meter   (5.11-1)

The A-Z P-Po Tab allows the unidirectional selection of the following dimensions:

Major/Minor-Tab   From/To Units on Pull-Down
A-Z / P-Po

parsec (pc) to meter (m)

peck (U.S.) (pk) to cubic meter (m^3)

peck (U.S.) (pk) to liter (L)

pennyweight (dwt) to kilogram (kg)

pennyweight (dwt) to gram (g)

perm (0 degC) to kilogram per pascal second square meter
[kg / (Pa ° s ° m^2)]

perm (23 degC) to kilogram per pascal second square meter
[kg / (Pa ° s ° m^2)]

perm inch (0 degC) to kilogram per pascal second meter
[kg / (Pa ° s ° m)]

perm inch (23 degC) to kilogram per pascal second meter
[kg / (Pa ° s ° m)]

phot (ph) to lux (lx)

pica (computer) (1/6 in) to meter (m)

Major/Minor-Tab   From/To Units on Pull-Down

A-Z / P-Po

pica (computer) (1/6 in) to millimeter (mm)

pica (printers) to meter (m)

pica (printers) to millimeter (mm)

pint (U.S. dry) (dry pt) to cubic meter (m^3)

pint (U.S. dry) (dry pt) to liter (L)

pint (U.S. liquid) (liq pt) to cubic meter (m^3)

pint (U.S. liquid) (liq pt) to liter (L)

point (computer) (1/72 in) to meter (m)

point (computer) (1/72 in) to millimeter (mm)

point (printers) to meter (m)

point (printers) to millimeter (mm)

poise (P) to pascal second (Pa * s)

pound (avoirdupois) (lb) to kilogram (kg)

pound (troy or apothecary) (lb) to kilogram (kg)

poundal to newton (N)

poundal per square foot to pascal (Pa)

poundal second per square foot to pascal second (Pa * s)

pound foot squared (lb * ft^2)

   to kilogram meter squared (kg * m^2)

pound-force (lbf) to newton (N)

pound-force foot (lbf * ft) to newton meter (N * m)

pound-force foot per inch (lbf * ft/in)

   to newton meter per meter (N * m/m)

5.12   PO-PZ TAB

The twelfth Minor Tab under the A-Z Major Tab is the Po-Pz Tab (see Figure 5.12-1). This example demonstrates the conversion of pound per gallon (U.S.) (lb/gal) to kilogram per cubic meter (kg/m^3). In this example, the amplitude of pound per gallon (U.S.) (lb/gal) is set equal to one (1.0). The conversion equation is shown in Example 5.12-1.

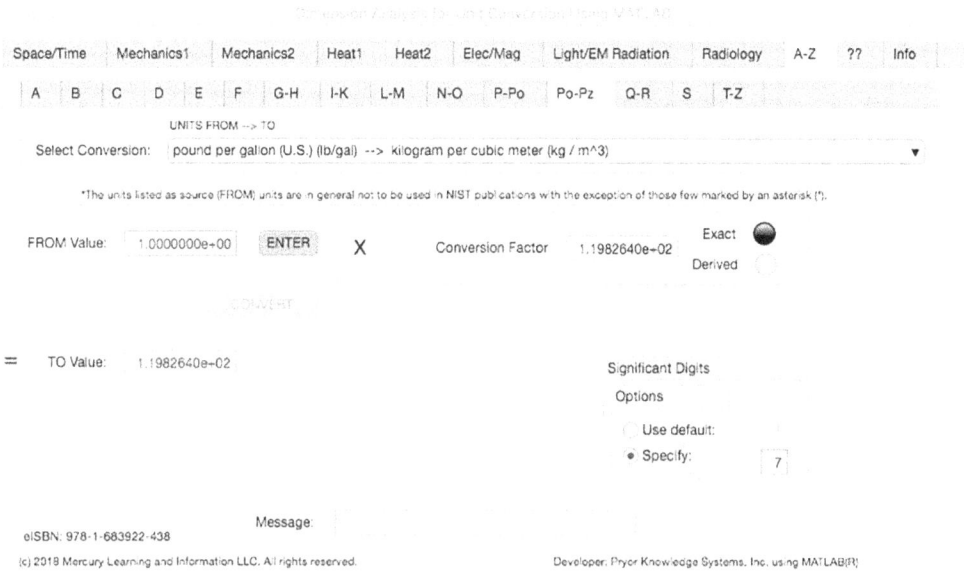

FIGURE 5.12-1 Unit Converter Using MATLAB App A-Z - Po-Pz Tab.

pound per gallon (U.S.) (lb/gal) (=1.0) to kilogram per cubic meter (kg/m^3)
Example 5.12-1:

$$kg/m^3 = lb/gal * kg/lb * gal/L * L/m^3$$
$$= lb/gal * 4.5359237E\text{-}1 * 2.6417204E\text{-}1 * 1.0E3$$
$$= lb/gal * 1.1982642E2$$

Where: kg/m^3 = the number of kilograms per cubic meter calculated
lb/gal = the number of pounds per gallon to be converted (1.0)
kg/lb = the conversion factor from pound to kilogram
(4.5359237E-1)
gal/L = the conversion factor from liters to gallon (2.6417204E-1)
L/m^3 = the conversion factor from cubic meters to liter (1.0E3)

And:

The resulting solution is:

m = 1.0 * 1.1982642E2
= 1.1982642E2 kilograms per cubic meter   (5.12-1)

The A-Z Po-Pz Tab allows the unidirectional selection of the following dimensions:

Major/Minor-Tab	From/To Units on Pull-Down
A-Z / Po-Pz	

pound-force inch (lbf * in) to newton meter (N * m)

pound-force inch per inch (lbf * in/in)
   to newton meter per meter (N * m/m)

pound-force per foot (lbf/ft) to newton per meter (N/m)

pound-force per inch (lbf/in) to newton per meter (N/m)

pound-force per pound (lbf/lb) (thrust to mass ratio)
   to newton per kilogram (N/kg)

pound-force per square foot (lbf/ft^2) to pascal (Pa)

pound-force per square inch (psi) (lbf/in^2) to pascal (Pa)

pound-force per square inch (psi) (lbf/in^2)
   to kilopascal (kPa)

pound-force second per square foot (lbf * s/ft^2)
   to pascal second (Pa * s)

pound-force second per square inch (lbf * s/in^2)
   to pascal second (Pa * s)

pound inch squared (lb * in^2)
   to kilogram meter squared (kg * m^2)

pound per cubic foot (lb/ft^3)
   to kilogram per cubic meter (kg/m^3)

pound per cubic inch (lb/in^3)
   to kilogram per cubic meter (kg/m^3)

pound per cubic yard (lb/yd^3)
   to kilogram per cubic meter (kg/m^3)

pound per foot (lb/ft) to kilogram per meter (kg/m)

pound per foot hour [lb/(ft * h)] to pascal second (Pa * s)

pound per foot second [lb/(ft * s)] to pascal second (Pa * s)

Major/Minor-Tab   From/To Units on Pull-Down

A-Z / Po-Pz

pound per gallon [Canadian and U.K. (Imperial)] (lb/gal)
    to kilogram per cubic meter (kg/m^3)

pound per gallon [Canadian and U.K. (Imperial)] (lb/gal)
    to kilogram per liter (kg/L)

pound per gallon (U.S.) (lb/gal)
    to kilogram per cubic meter (kg/m^3)

pound per gallon (U.S.) (lb/gal)
    to kilogram per liter (kg/L)

pound per horsepower hour [lb/(hp ° h)]
    to kilogram per joule (kg/J)

pound per hour (lb/h) to kilogram per second (kg/s)

pound per inch (lb/in) to kilogram per meter (kg/m)

pound per minute (lb/min) to kilogram per second (kg/s)

pound per second (lb/s) to kilogram per second (kg/s)

pound per square foot (lb/ft^2)
    to kilogram per square meter (kg/m^2)

pound per square inch (NOT pound-force) (lb/ft^2)
    to kilogram per square meter (kg/m^2)

pound per yard (lb/yd) to kilogram per meter (kg/m)

psi (pound-force per square inch) (lbf/in^2) to pascal (Pa)

psi (pound-force per square inch) (lbf/in^2)
    to kilopascal (kPa)

5.13   Q-R TAB

The thirteenth Minor Tab under the A-Z Major Tab is the Q-R Tab (see Figure 5.13-1). This example demonstrates the conversion of quart (U.S. liquid) (liq qt) (=1.0) to cubic meter (m^3). In this example, the amplitude of quart is set equal to one (1.0). The conversion equation is shown in Example 5.13-1.

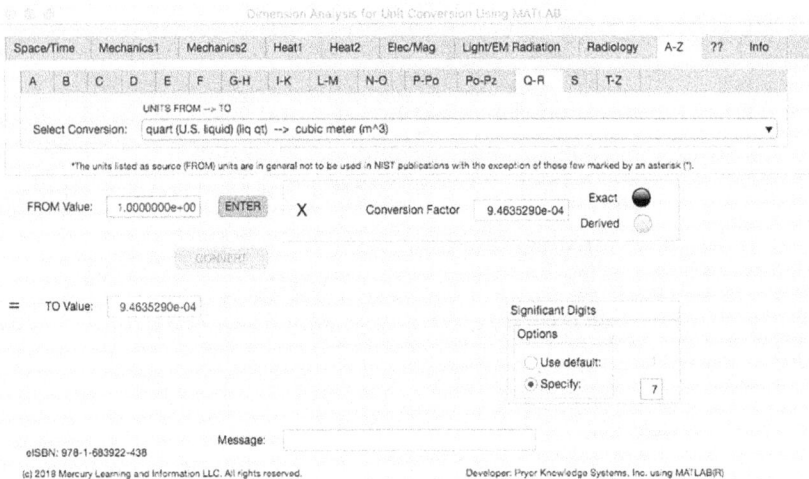

FIGURE 5.13-1 Unit Converter Using MATLAB App A-Z - Q-R Tab.

quart (U.S. liquid) (liq qt) (=1.0) to cubic meter (m^3) Example 5.13-1:

$$m^3 = \text{liq qt} * \text{L/liq qt} * m^3/\text{L}$$
$$= \text{liq qt} * = \text{liq qt} * 9.463529\text{E-1} * 1.0\text{E-3}$$
$$= \text{liq qt} * 9.463529\text{E-4}$$

Where: m^3 = the number of cubic meters calculated
   liq qt = the number of liquid quarts to be converted (1.0)
   L/liq qt = the conversion factor from liquid quart to liter (9.463529E-1)
   m^3/L = the conversion factor from liter to cubic meter (1.0E-3)

And:

   The resulting solution is:

   $$m^3 = 1.0 * 9.463529\text{E-4} = 9.463529\text{E-4} \text{ cubic meter} \tag{5.13-1}$$

The A-Z Q-R Tab allows the unidirectional selection of the following dimensions:

Major/Minor-Tab   From/To Units on Pull-Down
A-Z / Q-R

   quad (10^15 Btu IT) to joule (J)
   quart (U.S. dry) (dry qt) to cubic meter (m^3)
   quart (U.S. dry) (dry qt) to liter (L)
   quart (U.S. liquid) (liq qt) to cubic meter (m^3)

Major/Minor-Tab	From/To Units on Pull-Down
A-Z / Q-R	

quart (U.S. liquid) (liq qt) to liter (L)

°rad (absorbed dose) (rad) to gray (Gy)

°rem (rem) to sievert (Sv)

revolution (r) to radian (rad)

revolution per minute (rpm) (r / min)
 to radian per second (rad / s)

rhe to reciprocal pascal second (Pa ° s)^-1

rod (based on U.S. survey foot) (rd) to meter (m)

°roentgen (R) to coulomb per kilogram (C / kg)

rpm (revolution per minute) (r / min)
 to radian per second (rad / s)

5.14 S TAB

The fourteenth Minor Tab under the A-Z Major Tab is the S Tab (see Figure 5.14-1). This example demonstrates the conversion of square mile (mi^2) (=1.0) to square meter (m^2). In this example, the amplitude of square mile is set equal to one (1.0). The conversion equation is shown in Example 5.14-1.

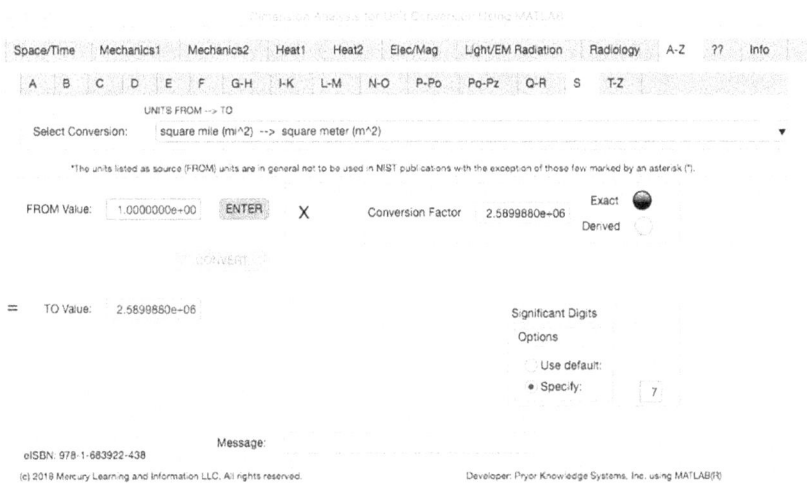

FIGURE 5.14-1 Unit Converter Using MATLAB App A-Z - S Tab.

square mile (mi^2) (=1.0) to square meter (m^2) Example 5.14-1:

m^2 = mi^2 * km^2/mi^2 * m^2/km^2
= mi^2 * 2.589988 * 1.0E6
= mi^2 * 2.589988E6

Where: m^2 = the number of square meters calculated
mi^2 = the number of square miles to be converted (1.0)
km^2/mi^2 = the conversion factor from square mile to square kilometer (2.589988)
m^2/km^2 = the conversion factor from square kilometer to square meter (1.0E6)

And:

The resulting solution is:

m^2 = 1.0 * 2.589988E6 = 2.589988E6 square meter   (5.14-1)

The A-Z S Tab allows the unidirectional selection of the following dimensions:

Major/Minor-Tab   From/To Units on Pull-Down
A-Z / S

*second (angle) (") to radian (rad)
second (sidereal) to second (s)
shake to second (s)
shake to nanosecond (ns)
slug (slug) to kilogram (kg)
slug per cubic foot (slug/ft^3)
   to kilogram per cubic meter (kg/m^3)
slug per foot second (slug/(ft * s)
   to pascal second (Pa * s)
square foot (ft^2) to square meter (m^2)
square foot per hour (ft^2/h)
   to square meter per second (m^2/s)
square foot per second (ft^2/s)
   to square meter per second (m^2/s)

Major/Minor-Tab	**From/To Units on Pull-Down**
A-Z / S	

square inch (in^2) to square meter (m^2)

square inch (in^2) to square centimeter (cm^2)

square mile (mi^2) to square meter (m^2)

square mile (mi^2) to square kilometer (km^2)

square mile (based on U.S. survey foot) (mi^2) to square meter (m^2)

square mile (based on U.S. survey foot) (mi^2) to square kilometer (km^2)

square yard (yd^2) to square meter (m^2)

statampere to ampere (A)

statcoulomb to coulomb (C)

statfarad to farad (F)

stathenry to henry (H)

statmho to siemens (S)

statohm to ohm (ohm)

statvolt to volt (V)

stere (st) to cubic meter (m^3)

stilb (sb) to candela per square meter (cd / m^2)

stokes (St) to meter squared per second (m^2 / s)

5.15   T-Z TAB

The fifteenth Minor Tab under the A-Z Major Tab is the T-Z Tab (see Figure 5.15-1). This example demonstrates the conversion of year (365 days) (=1.0) to second (s). In this example, the amplitude of year is set equal to one (1.0). The conversion equation is shown in Example 5.15-1 {37}.

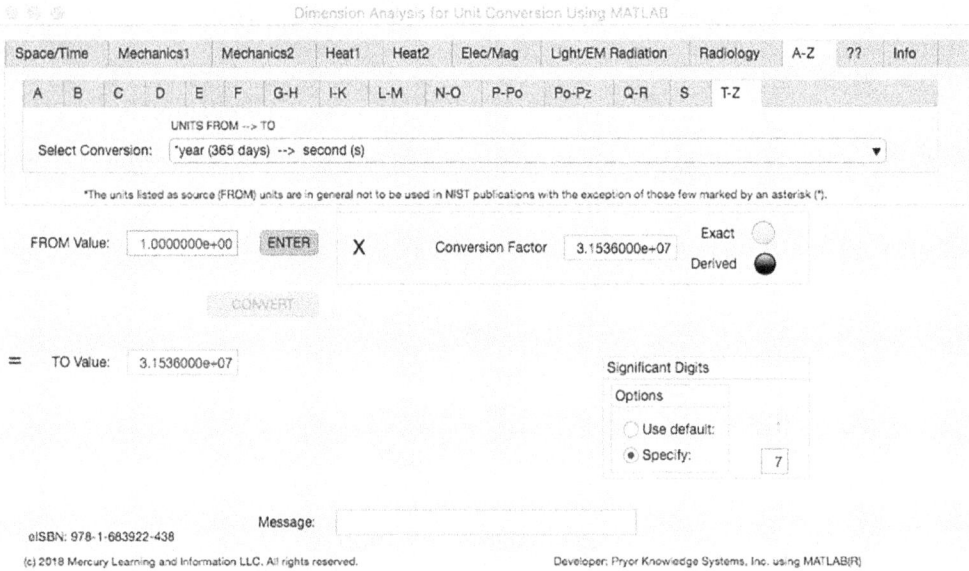

FIGURE 5.15-1 Unit Converter Using MATLAB App A-Z - T-Z Tab.

year (365 days) (=1.0) to second (s) Example 5.15-1:

s = year (365 days) * d/year (365 days) * h/d * min/h * s/min
= 1.0 * 365 * 24 * 60 * 60 = 3.1536E7

Where: s = the number of seconds calculated
year = the number of years to be converted (1.0)
d/year = the conversion factor from year to day (365)
h/d = the conversion factor from day to hour (24)
min/h = the conversion factor from hour to minute (60)
s/min = the conversion factor from minute to second (60)

And:

The resulting solution is:

s = 1.0 * 3.1536E7 = 3.1536E7 second per 365 day year   (5.15-1)

The A-Z T-Z Tab allows the unidirectional selection of the following dimensions:

Major/Minor-Tab   From/To Units on Pull-Down
A-Z / T-Z

tablespoon to cubic meter (m^3)

tablespoon to milliliter (mL)

teaspoon to cubic meter (m^3)

teaspoon to milliliter (mL)

tex to kilogram per meter (kg/m)

therm (EC) to joule (J)

therm (U.S.) to joule (J)

ton, assay (AT) to kilogram (kg)

ton, assay (AT) to gram (g)

ton-force (2000 lbf) to newton (N)

ton-force (2000 lbf) to kilonewton (kN)

ton, long (2240 lb) to kilogram (kg)

ton, long, per cubic yard
    to kilogram per cubic meter (kg/m^3)

*ton, metric to kilogram (kg)

tonne (called metric ton in U.S.) (t) to kilogram (kg)

ton of refrigeration (12 000 Btu IT / h) to watt (W)

ton of TNT (energy equivalent) to joule (J)

ton, register to cubic meter (m^3)

ton, short (2000 lb) to kilogram (kg)

ton, short, per cubic yard
    to kilogram per cubic meter (kg/m^3)

ton, short, per hour to kilogram per second (kg/s)

torr (Torr) to pascal (Pa)

unit pole to weber (Wb)

*watt hour (W * h) to joule (J)

*watt per square centimeter (W/cm^2)
    to watt per square meter (W/m^2)

Major/Minor-Tab   **From/To Units on Pull-Down**
A-Z / T-Z

watt per square inch (W/in^2)
    to watt per square meter (W/m^2)
*watt second (W * s) to joule (J)
yard (yd) to meter (m)
*year (365 days) to second (s)
year (sidereal) to second (s)
year (tropical) to second (s)

REFERENCES

1. Webster's New World Dictionary, Second College Edition, Simon and Schuster, New York, p 396
2. Webster's New World Dictionary, Second College Edition, Simon and Schuster, New York, p 1489
3. Webster's New World Dictionary, Second College Edition, Simon and Schuster, New York, p 361
4. Webster's New World Dictionary, Second College Edition, Simon and Schuster, New York, p 960
5. Webster's New World Dictionary, Second College Edition, Simon and Schuster, New York, p 817
6. Webster's New World Dictionary, Second College Edition, Simon and Schuster, New York, p 923
7. Webster's New World Dictionary, Second College Edition, Simon and Schuster, New York, p 1389
8. Webster's New World Dictionary, Second College Edition, Simon and Schuster, New York, p 1088
9. Webster's New World Dictionary, Second College Edition, Simon and Schuster, New York, p 894
10. Webster's New World Dictionary, Second College Edition, Simon and Schuster, New York, p 283
11. Webster's New World Dictionary, Second College Edition, Simon and Schuster, New York, p 842
12. Webster's New World Dictionary, Second College Edition, Simon and Schuster, New York, p 1464
13. Webster's New World Dictionary, Second College Edition, Simon and Schuster, New York, p 1426

14. Webster's New World Dictionary, Second College Edition, Simon and Schuster, New York, p 1631

15. Webster's New World Dictionary, Second College Edition, Simon and Schuster, New York, p 1647

16. Webster's New World Dictionary, Second College Edition, Simon and Schuster, New York, p 1285

17. Webster's New World Dictionary, Second College Edition, Simon and Schuster, New York, p 525

18. Webster's New World Dictionary, Second College Edition, Simon and Schuster, New York, p 269

19. Webster's New World Dictionary, Second College Edition, Simon and Schuster, New York, p 1312

20. Webster's New World Dictionary, Second College Edition, Simon and Schuster, New York, p 202

21. Webster's New World Dictionary, Second College Edition, Simon and Schuster, New York, p 1642

22. Webster's New World Dictionary, Second College Edition, Simon and Schuster, New York, p 323

23. Webster's New World Dictionary, Second College Edition, Simon and Schuster, New York, p 312

24. Webster's New World Dictionary, Second College Edition, Simon and Schuster, New York, p 792

25. Jonathan Haidt, *The Righteous Mind*, Vintage Books, Random House Inc., New York, ISBN 978-0-307-45577-2, p 54

26. Webster's New World Dictionary, Second College Edition, Simon and Schuster, New York, p 1179

27. Webster's New World Dictionary, Second College Edition, Simon and Schuster, New York, p 1478

28. J. D. Barrow, *The Constants of Nature*, Pantheon Books, New York, ISBN 0-375-42221-8, p 47

29. J. D. Barrow and F. J. Tipler, *The Anthropic Cosmological Principle*, Oxford University Press, Oxford New York, ISBN 0-19-282147-4, pp 458-509

30. K. Thompson, and B. N. Taylor, *Guide for the Use of the International System of Units (SI)*, Natl. Inst. Stand. Technol. Spec. Pub. 811, 24 Pages (May 2006)

31. A. Butcher, L. Crown and E. J. Gentry, *The International System of Units (SI) – Conversion Factors for General Use*, Natl. Inst. Stand. Technol. Spec. Pub. 1038, 78 Pages (March 2008)

32. Webster's New World Dictionary, Second College Edition, Simon and Schuster, New York, p 473

33. Webster's New World Dictionary, Second College Edition, Simon and Schuster, New York, p 1210

34. Webster's New World Dictionary, Second College Edition, Simon and Schuster, New York, p 1552

35. Webster's New World Dictionary, Second College Edition, Simon and Schuster, New York, p 276

36. A. Butcher, L. Crown and E. J. Gentry, *The International System of Units (SI) – Conversion Factors for General Use*, Natl. Inst. Stand. Technol. Spec. Pub. 1038, (March 2008), pp 8-15

37. Conversion Factors shown in this equation in BOLD font are defined to be exact. Conversion Factors not shown in this equation in BOLD font are defined as derived. All later conversion equations will not make these distinctions. Please refer to reference documents {30,31} for detailed definitions on conversion factors.

38. The Liter, equal to 0.001 m^3, is accepted for use with SI. A. Butcher, L. Crown and E. J. Gentry, *The International System of Units (SI) – Conversion Factors for General Use*, Natl. Inst. Stand. Technol. Spec. Pub. 1038, (March 2008), p 9

A

INSTALLATION, QUICK START, AND USE OF THE ENHANCED UNIT CONVERTER MATLAB APP

SYSTEM REQUIREMENTS

This App is designed to run under all operating systems that have a running installed copy of MATLAB Versions: R2017b or R2018a. For compatible systems, go to *https://www.mathworks.com/support/sysreq.html*.

INSTALLATION

Download the installation file as directed by the publisher.

Open the installation file to store the installation file, Enhanced Unit Converter Using MATLAB App.mlappinstall on your computer.

Start MATLAB. Select the APPS Tab. Click the Install App button, navigate to the installation file on disk, select the file, and click the Open button.

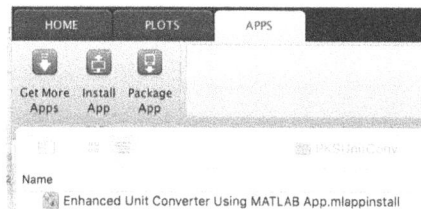

FIGURE A.1 Installation File Selection.

MATLAB will ask you to confirm the installation with a message window. Click the Install button to confirm.

FIGURE A.2 Confirm Installation.

The app will be installed and appear in the toolbar to the right of the currently installed app buttons.

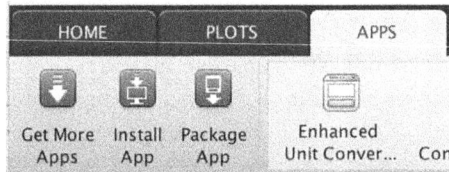

FIGURE A.3 App Toolbar with the Installed App.

APP DETAILS

OVERVIEW

This app converts values from one unit of measure to another using standard conversion factors. It performs conversions from and to the inch-pound system units used in the United States of America and the International System of Units (SI) as documented in the National Institute of Standards and Technology (NIST) publications of conversions for general use. [1,2] There are 1,316 conversion factors available for bidirectional conversion from/to SI units, organized into 44 minor subsections by topic under eight major topical sections. There is also an alphabetical section comprising 445 conversion factors for unidirectional conversion to SI units.

FIGURE B.1 The Unit Converter Application Window.

The application performs all three steps in the conversion process: application of the relevant conversion factor, selection of significant digits, and rounding of the result. Conversion factors designated as "exact" are definitions, or they have been set by agreements that define the factor value precisely. All other conversion factors, designated as "derived," result from truncation of decimal places and/or calculation by a combination of other factors.

INPUT VARIABLES

Units of measure

a. Select the major category of units to be converted using the tab group at the top of the application window.

b. Select the minor category of units to be converted using the lower tab group.

c. Use the Left pull-down list to select units to be converted (From) and use the Right pull-down list to select units to be converted (To).

The available categories are:

Major	Minor	Units
Space/Time		
	Length	
		nautical mile (nmi)
		mile (mi)
		kilometer (km)
		fathom (ftm)
		yard (yd)
		meter (m)
		foot (ft)
		foot (U.S. Survey)
		inch (in)
		centimeter (cm)
		millimeter (mm)
		pica, printer's (12p)
		point, printer's (p)

Major	Minor	Units
Space/Time		
	Length	
		mil (0.001 in)
		micrometer (μm)
		microinch (μin)
		nanometer (nm)
		angstrom
	Plane Angle	
		Radian
		Degree Arc
	Area	
		square mi (mi^2)
		square km (km^2)
		hectare (ha)
		acre
		square m (m^2)
		square yd (yd^2)
		square ft (ft^2)
		square in (in^2)
		square cm (cm^2)
		square mm (mm^2)
		circular mil
	Volume	
		acre-foot (aft)
		barrel-oil (bbl)
		cubic-yard (yd^3)
		cubic-meter (m^3)
		cubic-foot (ft^3)
		board-foot (BF)
		register-ton (RT)
		bushel (bu)
		gallon (gal)

Major	Minor	Units
Space/Time	Volume	
		liter (L)
		quart (qt)
		pint (pt)
		fluid-ounce (fl oz)
		milliliter (mL)
		cubic-inch (in^3)
		cubic-centimeter (cm^3)
	Velocity	
		foot per second (ft/s)
		mile per hour (mi/hr)
		knot (nmi/hr)
		meter per second (m/s)
		kilometer per hour (km/hr)
	Acceleration	
		inch per second squared (in/s^2)
		foot per second squared (ft/s^2)
		meter per second squared (m/s^2)
		standard acceleration of gravity (g)
	Flow rate	
		cubic foot per second (ft^3/s)
		cubic foot per minute (ft^3/min)
		cubic yard per minute (yd^3/min)
		gallon per minute (gal/min)
		gallon per day (gal/day)
		cubic meter per second (m^3/s)
		liter per second (L/s)
		liter per day (L/day)
	Fuel efficiency	
		miles per gallon (mi/gal)
		kilometer per liter (km/L)

Major	Minor	Units
Mechanics 1		
	Mass (Weight)	
		ton (long) (2240 lb)
		metric ton (t)
		ton (short) (2000 lb)
		slug
		kilogram (kg)
		pound (avoirdupois)
		ounce (troy)
		ounce (avoirdupois)
		gram (g)
		grain
		milligram (mg)
	Moment of Mass	
		pound foot
		kilogram meter (kg ° m)
	Density	
		ton (2000 lb [short]) per cubic yard
		metric ton per cubic meter (t/m^2)
		pound per cubic foot (lb/ft^3)
		kilogram per cubic meter (kg/m^3)
	Concentration (Mass)	
		pound per gallon (lb/gal)
		ounce (avoirdupois) per gallon (oz/gal)
		gram per liter (g/L)
	Momentum	
		pound foot per second (lb°ft/s)
		kilogram meter per second (kg°m/s)
	Moment of Inertia	
		pound square foot (lb°ft^2)
		kilogram square meter (kg°m^2)
	Force	
		pound-force
		poundal
		newton (N)

Major	Minor	Units
Mechanics 2		
	Moment of Force, Torque	
		pound-force foot
		pound-force inch
		newton meter (N*m)
	Pressure, Stress	
		megapascal (Mpa)
		standard atmosphere
		bar
		kilopascal (kPa)
		millibar
		pound-force per square inch (psi)
		kilopound-force per square inch
		pound-force per square foot
		inch of mercury (32 degF)
		foot of water (39.2 degF)
		inch of water (39.2 degF)
		millimeter of mercury (32 degF)
		torr (Torr)
		pascal (Pa)
	Viscosity (Dynamic)	
		centipoise
		millipascal second (mPa*s)
	Viscosity (Kinematic)	
		centistokes
		square millimeter per second (mm^2/s)

Major	Minor	Units
Mechanics 2		
	Energy, Work, Heat	
		kilowatthour
		megajoule (MJ)
		calorie (physics)
		kilojoule (kJ)
		calorie (nutrition) (kCal)
		joule (J)
		Btu
		therm (U.S.)
		horsepower hour
		foot pound-force
	Power	
		ton, refrigeration (12 000 Btu/h)
		kilowatt (kW)
		Btu per second
		Btu per hour
		watt (W)
		horsepower (550 ft-lbF/s)
		horsepower, electric
		foot pound-force per second (ft-lbF/s)
Heat 1		
	Temperature	
		Fahrenheit
		Celsius
		kelvin
	Linear Expansion Coefficient	
		deltaT Fahrenheit (deltaF)
		deltaT kelvin (deltaK)
		deltaT Celsius (deltaC)

Major	Minor	Units
Heat 1		
	Thermal Conductivity	
		Btu inch per hour square foot degree Fahrenheit
		watt per meter kelvin [W/(m*K)]
	Coefficient of Heat Transfer	
		Btu per hour square foot degree Fahrenheit
		watt per square meter kelvin [W/(m^2*K)]
	Heat Capacity	
		Btu per degree Fahrenheit
		kilojoule per kelvin (kJ/K)
Heat 2		
	Specific Heat Capacity	
		Btu per pound degree Fahrenheit
		kilojoule per kilogram kelvin [kJ/(kg*K)]
	Entropy	
		Btu per degree Rankine
		kilojoule per kelvin [kJ/(K)]
	Specific Entropy	
		Btu per pound degree Rankine
		kilojoule per kilogram kelvin [kJ/(kg*K)]
	Specific Internal Energy	
		Btu per pound

Major	Minor	Units
		kilojoule per kilogram [kJ/kg]
Electricity/ Magnetism		
	Magnetic Field Strength	
		oersted
		ampere per meter (A/m)
	Magnetic Flux	
		maxwell
		nanoweber (nWb)
	Magnetic Flux Density	
		gauss
		millitesla (mT)
	Electric Charge	
		ampere hour
		coulomb (C)
	Resistivity	
		ohm circular mil per foot
		nanoohm meter (nOhm * m)
	Conductivity	
		mho per centimeter
		siemens per meter (S/m)
Light/EM Radiation		
	Wavelength	
		angstrom
		nanometer (nm)
	Luminance	
		lambert (L)
		candela per square inch
		footlambert
		candela per square meter (cd/m^2)

Major	Minor	Units
Light/EM Radiation		
	Luminous Excitance	
		lumen per square foot
		phot
		lux (lx)
	Illuminance	
		footcandle
		lux (lx)
Radiology		
	Activity of a Radionuclide	
		Curie
		megabecquerel (MBq)
	Absorbed Dose	
		Rad
		gray (Gy)
		centigray (cGy)
	Dose Equivalent	
		Rem
		millirem
		sievert
		millisievert
		microsievert
	Exposure (x and gamma rays)	
		roentgen
		coulomb per kilogram (C/kg)

Major/Minor Tab   Unidirectional Conversion Factors
A-Z / A

abampere to ampere (A)

abcoulomb to coulomb (C)

abfarad to farad (F)

abhenry to henry (H)

abmho to siemens (S)

abohm to ohm (Ω)

abvolt to volt (V)

acceleration in free fall, standard (g)
   to meter per second squared (m/s^2)

acre (based on U.S. Survey foot) to square meter (m^2)

acre foot (based on U.S. survey foot to cubic meter (m^3)

*ampere hour (a * h) to coulomb (C)

angstrom (Å) to meter (m)

angstrom (Å) to nanometer (nm)

are (a) to square meter

*astronomical unit (ua) to meter (m)

atmosphere, standard (atm) to pascal (Pa)

atmosphere, standard (atm) to kilopascal (kPa)

atmosphere, technical (at) to pascal (Pa)

atmosphere, technical (at) to kilopascal (kPa)

A-Z / B

*bar (bar) to pascal (Pa)

*bar (bar) to kilopascal (kPa)

*barn (b) to square meter (m^2)

barrel [for petroleum, 42 gallons (US)] (bbl)
   to cubic meter (m^3)

barrel [for petroleum, 42 gallons (US)] (bbl) to liter (L)

biot (Bi) to ampere (A)

British thermal unit IT (Btu IT) to joule (J)

Major/Minor Tab **Unidirectional Conversion Factors**
A-Z / B

British thermal unit th (Btu IT) to joule (J)

British thermal unit (mean) (Btu) to joule (J)

British thermal unit (39 degF) to joule (J)

British thermal unit (59 degF) (Btu) to joule (J)

British thermal unit (60 degF) (Btu) to joule (J)

Btu IT foot per hour square foot degree Fahrenheit
[Btu IT°ft/(h°ft^2°degF)]
to watt per meter kelvin [W/(m ° K)]

Btu th foot per hour square foot degree Fahrenheit
[Btu IT°ft/(h°ft^2°degF)]
to watt per meter kelvin [W/(m ° K)]

Btu IT inch per hour square foot degree Fahrenheit
[Btu IT°in/(h°ft^2°degF)]
to watt per meter kelvin [W/(m ° K)]

Btu th inch per hour square foot degree Fahrenheit
[Btu th°in/(h°ft^2°degF)]
to watt per meter kelvin [W/(m ° K)]

Btu IT inch per second square foot degree Fahrenheit
[Btu IT°in/(s°ft^2°degF)]
to watt per meter kelvin [W/(m ° K)]

Btu th inch per second square foot degree Fahrenheit
[Btu th°in/(s°ft^2°degF)]
to watt per meter kelvin [W/(m ° K)]

Btu IT per cubic foot (Btu IT/ft^3)
to joule per cubic meter (J/m^3)

Btu th per cubic foot (Btu th/ft^3)
to joule per cubic meter (J/m^3)

Btu IT per degree Fahrenheit (Btu IT/degF)
to joule per kelvin (J/K)

Major/Minor Tab **Unidirectional Conversion Factors**
A-Z / B

Btu th per degree Fahrenheit (Btu th/degF)
   to joule per kelvin (J/K)
Btu IT per degree Rankine (Btu IT/degR)
   to joule per kelvin (J/K)
Btu th per degree Rankine (Btu th/degR)
   to joule per kelvin (J/K)
Btu IT per hour (Btu IT/h) to Watt (W)
Btu th per hour (Btu th/h) to Watt (W)
Btu IT per hour square foot degree Fahrenheit
                    [Btu IT/(h°ft^2°degF)]
   to watt per square meter kelvin [W/(m^2° K)]
Btu th per hour square foot degree Fahrenheit
                    [Btu th/(h°ft^2°degF)]
   to watt per square meter kelvin [W/(m^2° K)]
Btu th per minute (Btu th/min) to watt (W)
Btu IT per pound (Btu IT/lb) to joule per kilogram (J/kg)
Btu th per pound (Btu th /lb) to joule per kilogram (J/kg)
Btu IT per pound degree Fahrenheit [Btu IT/(lb°degF)]
   to joule per kilogram kelvin [J/(kg°K)]
Btu th per pound degree Fahrenheit [Btu th/(lb°degF)]
   to joule per kilogram kelvin [J/(kg°K)]
Btu IT per pound degree Rankine [Btu IT/(lb°degR)]
   to joule per kilogram kelvin [J/(kg°K)]
Btu th per pound degree Rankine [Btu th/(lb°degR)]
   to joule per kilogram kelvin [J/(kg°K)]
Btu IT per second (Btu IT/s) to watt (W)
Btu th per second (Btu th/s) to watt (W)
Btu IT per second square foot degree Fahrenheit
                    [Btu IT/(s°ft^2°degF)]

Major/Minor Tab   **Unidirectional Conversion Factors**

A-Z / B

               to watt per square meter kelvin [W/(m^2*K)]

      Btu th per second square foot degree Fahrenheit

                                [Btu th/(s*ft^2*degF)]

               to watt per square meter kelvin [W/(m^2*K)]

      Btu IT per square foot [Btu IT/(ft^2)]

               to joule per square meter [J/(m^2)]

      Btu th per square foot [Btu th/(ft^2)]

               to joule per square meter [J/(m^2)]

      Btu IT per square foot hour [Btu IT/(ft^2*h)]

               to watt per square meter [W/(m^2)]

      Btu th per square foot hour [Btu th/(ft^2*h)]

               to watt per square meter [W/(m^2)]

      Btu th per square foot minute [Btu th/(ft^2*min)]

               to watt per square meter [W/(m^2)]

      Btu IT per square foot second [Btu IT/(ft^2*s)]

               to watt per square meter [W/(m^2)]

      Btu th per square foot second [Btu th/(ft^2*s)]

               to watt per square meter [W/(m^2)]

      Btu th per square inch second [Btu th/(in^2*s)]

               to watt per square meter [W/(m^2)]

      bushel (US) (bu) to cubic meter (m^3)

      bushel (US) (bu) to liter (L)

A-Z / C

      calorie$_{IT}$ (cal$_{IT}$) to joule (J)

      calorie$_{th}$ (cal$_{th}$) to joule (J)

      calorie (cal) (mean) to joule (J)

      calorie (15 °C) (cal) to joule (J)

      calorie (20 °C) (cal) to joule (J)

      calorie$_{IT}$, kilogram (nutrition) to joule (J)

Major/Minor Tab   **Unidirectional Conversion Factors**
A-Z / C

calorie$_{th}$, kilogram (nutrition) to joule (J)

calorie (mean), kilogram (nutrition) to joule (J)

calorie$_{th}$ per centimeter second degree Celsius

$$[cal_{th}/(cm^\circ s^{\circ\circ}C)]$$

to watt per meter kelvin [W/(m$^\circ$K)]

calorie$_{IT}$ per gram (cal$_{IT}$/g) to joule per kilogram (J/kg)

calorie$_{th}$ per gram (cal$_{th}$/g) to joule per kilogram (J/kg)

calorie$_{IT}$ per gram degree Celsius [cal$_{IT}$/(g$^{\circ\circ}$C)]

to joule per kilogram kelvin [J/(kg$^\circ$K)]

calorie$_{th}$ per gram degree Celsius [cal$_{th}$/(g$^{\circ\circ}$C)]

to joule per kilogram kelvin [J/(kg$^\circ$K)]

calorie$_{IT}$ per gram kelvin [cal$_{IT}$/(g$^\circ$K)]

to joule per kilogram kelvin [J/(kg$^\circ$K)]

calorie$_{th}$ per gram kelvin [cal$_{th}$/(g$^\circ$K)]

to joule per kilogram kelvin [J/(kg$^\circ$K)]

calorie$_{th}$ per minute (cal$_{th}$/min) to watt (W)

calorie$_{th}$ per second (cal$_{th}$/s) to watt (W)

calorie$_{th}$ per square centimeter (cal$_{th}$/cm^2)

to joule per square meter (J/m^2)

calorie$_{th}$ per square centimeter minute [cal$_{th}$/(cm^2°min)]

to watt per square meter (watt/m^2)

calorie$_{th}$ per square centimeter second [cal$_{th}$/(cm^2 $^\circ$ s)]

to watt per square meter (watt/m^2)

candela per square inch (cd/in^2)

to candela per square meter (cd/m^2)

carat, metric to kilogram (kg)

carat, metric to gram (g)

centimeter of mercury (0 degC) to pascal (Pa)

centimeter of mercury (0 degC) to kilopascal (kPa)

Major/Minor Tab
A-Z / C

Unidirectional Conversion Factors

centimeter of mercury, conventional (cmHg) to pascal (Pa)

centimeter of mercury, conventional (cmHg)
   to kilopascal (kPa)

centimeter of water (4 degC) to pascal (Pa)

centimeter of water, conventional (cmH2O) to pascal (Pa)

centipoise (cP) to pascal second (Pa * s)

centistokes (cSt) to meter squared per second (m^2 / s)

chain (based on US survey foot) (ch) to meter (m)

circular mil to square meter (m^2)

circular mil to square millimeter (mm^2)

clo to square meter kelvin per watt (m^2 * K / W)

cord (128 ft^3) to cubic meter (m^3)

cubic foot (ft^3) to cubic meter (m^3)

cubic foot per minute (ft^3 / min)
   to cubic meter per second (m^3/s)

cubic foot per minute (ft^3 / min) to liter per second (L / s)

cubic foot per second (ft^3 / s)
   to cubic meter per second (m^3/s)

cubic inch (in^3) to cubic meter (m^3)

cubic inch per minute (in^3/min)
   to cubic meter per second (m^3/s)

cubic mile (mi^3) to cubic meter (m^3)

cubic yard (yd^3) to cubic meter (m^3)

cubic yard per minute (yd^3/min)
   to cubic meter per second (m^3/s)

cup (US) to cubic meter (m^3)

cup (US) to liter (L)

cup (US) to milliliter (mL)

*Curie (Ci) to becquerel (Bq)

Major/Minor Tab   **Unidirectional Conversion Factors**
A-Z/D

darcy to meter squared (m^2)

*day (d) to second (s)

day (sidereal) to second (s)

debye (D) to coulomb meter (C * m)

*degree (angle) to radian (rad)

*degree Celsius (temperature interval) (degC) to kelvin (K)

degree centigrade (temperature interval)
   to degree Celsius (degC)

degree Fahrenheit (temperature interval) (degF)
   to degree Celsius (degC)

degree Fahrenheit (temperature interval) (degF)
   to kelvin (K)

degree Fahrenheit hour per British thermal unit IT
                   (degF * h / Btu IT)
   to kelvin per watt (K / W)

degree Fahrenheit hour per British thermal unit th
                   (degF * h / Btu th)
   to kelvin per watt (K / W)

degree Fahrenheit hour square foot per Btu IT
                 (degF*h*ft^2 / Btu IT)
   to square meter kelvin per watt (m^2*K / W)

degree Fahrenheit hour square foot per Btu th
                 (degF*h*ft^2 / Btu th)
   to square meter kelvin per watt (m^2*K / W)

degree Fahrenheit hour square foot per Btu IT inch
                 (degF*h*ft^2 / Btu IT*in)
   to meter kelvin per watt (m*K / W)

degree Fahrenheit hour square foot per Btu th inch
                 (degF*h*ft^2 / Btu th*in)

Major/Minor Tab **Unidirectional Conversion Factors**

A-Z/D

    to meter kelvin per watt (m°K / W)

degree Fahrenheit second per Btu IT (degF°s / Btu IT)
    to kelvin per watt (K / W)

degree Fahrenheit second per Btu th (degF°s / Btu th)
    to kelvin per watt (K / W)

degree Rankine (temperature interval) (degR) to kelvin (K)

denier to kilogram per meter (kg / m)

denier to gram per meter (g / m)

dyne (dyn) to newton (N)

dyne centimeter (dyn-cm) to newton meter (N ° m)

dyne per square centimeter (dyn / cm^2) to pascal (Pa)

A-Z/E

°electronvolt (eV) to joule (J)

EMU of capacitance (abfarad) to farad (F)

EMU of current (abampere) to ampere (A)

EMU of Electric potential (abvolt) to volt (V)

EMU of inductance (abhenry) to henry (H)

EMU of resistance (abohm) to ohm

erg (erg) to joule (J)

erg per second (erg / s) to watt (W)

erg per square centimeter second [erg / (cm^2 ° s)]
    to watt per square meter (W / m^2)

ESU of capacitance (statfarad) to farad (F)

ESU of current (statampere) to ampere (A)

ESU of Electric potential (statvolt) to volt (V)

ESU of inductance (stathenry) to henry (H)

ESU of resistance (statohm) to ohm

A-Z/F

faraday (based on carbon 12) to coulomb (C)

Major/Minor Tab   **Unidirectional Conversion Factors**
A-Z/F

       fathom (based on US survey foot) to meter (m))

       fermi to meter (m)

       fermi to femtometer (fm)

       fluid ounce (US) (fl oz) to cubic meter (m^3)

       fluid ounce (US) (fl oz) to milliliter (mL)

       foot (ft) to meter (m)

       foot (US Survey) (ft) to meter (m)

       footcandle to lux (lx)

       footlambert to candela per square meter (cd / m^2)

       foot of mercury, conventional (ftHg) to pascal (Pa)

       foot of mercury, conventional (ftHg) to kilopascal (kPa)

       foot of water (39.2 degF) to pascal (Pa)

       foot of water (39.2 degF) to kilopascal (kPa)

       foot of water, conventional (ftH2O) to pascal (Pa)

       foot of water, conventional (ftH2O) to kilopascal (kPa)

       foot per hour (ft /h) to meter per second (m / s)

       foot per minute (ft /min) to meter per second (m / s)

       foot per second (ft/s) to meter per second (m / s)

       foot per second squared (ft/s^2)
         to meter per second squared (m/s^2)

       foot poundal to joule (J)

       foot pound-force (ft ° lbf) to joule (J)

       foot pound-force per hour (ft ° lbf/h) to watt (W)

       foot pound-force per minute (ft ° lbf/min) to watt (W)

       foot pound-force per second (ft ° lbf/s) to watt (W)

       foot to the fourth power (ft^4)
         to meter to the fourth power (m^4)

       franklin (Fr) to coulomb (C)

Major/Minor Tab
A-Z/G-H

Unidirectional Conversion Factors

gal (Gal) to meter per second squared (m / s^2)

gallon [Canadian and U.K. (Imperial)] (gal)
   to cubic meter (m^3)

gallon [Canadian and U.K. (Imperial)] (gal) to liter (L)

gallon (US) (gal) to cubic meter (m^3)

gallon (US) (gal) to liter (L)

gallon (US) per day (gal/d)
   to cubic meter per second (m^3/s)

gallon (US) per day (gal/d) to liter per second (L/s)

gallon (US) per horsepower hour [gal/(hp ° h)]
   to cubic meter per joule (m^3/J)

gallon (US) per horsepower hour [gal/(hp ° h)]
   to liter per joule (L/J)

gallon (US) per minute (gpm) (gal/min)
   to cubic meter per second (m^3/s)

gallon (US) per minute (gpm) (gal/min)
   to liter per second (L/s)

gammato tesla (T)

gauss (Gs, G) to tesla (T)

gilbert (Gi) to ampere (A)

gill [Canadian and U.K. (Imperial)] (gi)
   to cubic meter (m^3)

gill [Canadian and U.K. (Imperial)] (gi) to liter (L)

gill (US) (gi) to cubic meter (m^3)

gill (US) (gi) to liter (L)

gon (also called grade) (gon) to radian (rad)

gon (also called grade) (gon) to degree (angle)

grain (gr) to kilogram (kg)

grain (gr) to milligram (mg)

Major/Minor Tab **Unidirectional Conversion Factors**
A-Z/G-H

grain per gallon (US) (gr/gal)
   to kilogram per cubic meter (kg/m^3)

grain per gallon (US) (gr/gal)
   to milligram per liter (mg/L)

gram-force per square centimeter (gf/cm^2) to pascal (Pa)

*gram per cubic centimeter (g/cm^3)
   to kilogram per cubic meter (kg/m^3)

*hectare (ha) to square meter (m^2)

horsepower (550 ft * lbf / s) (hp) to watt (W)

horsepower (boiler) to watt (W)

horsepower (electric) to watt (W)

horsepower (metric) to watt (W)

horsepower (U.K.) to watt (W)

horsepower (water) to watt (W)

*hour to second (s)

hour (sidereal) to second (s)

hundredweight (long, 112 lb) to kilogram (kg)

hundredweight (short, 100 lb) to kilogram (kg)

A-Z/I-K

inch (in) to meter (m)

inch (in) to centimeter (cm)

inch of mercury (32 degF) to pascal (Pa)

inch of mercury (32 degF) to kilopascal (kPa)

inch of mercury (60 degF) to pascal (Pa)

inch of mercury (60 degF) to kilopascal (kPa)

inch of mercury, conventional (inHg) to pascal (Pa)

inch of mercury, conventional (inHg) to kilopascal (kPa)

inch of water (39.2 degF) to pascal (Pa)

inch of water (60 degF) to pascal (Pa)

Major/Minor Tab **Unidirectional Conversion Factors**
A-Z/I-K

inch of water, conventional (in H2O) to pascal (Pa)

inch per second (in/s) to meter per second (m / s)

inch per second squared (in/s^2)
   to meter per second squared (m/s^2)

inch to the fourth power (in^4)
   to meter to the fourth power (m^4)

kayser (K) to reciprocal meter (m^-1)

kilocalorie IT (kcal IT) to joule (J)

kilocalorie th (kcal th) to joule (J)

kilocalorie (mean) (kcal) to joule (J)

kilocalorie th per minute (kcal th / min) to watt (W)

kilocalorie th per second (kcal th / s) to watt (W)

kilogram-force (kgf) to newton (N)

kilogram-force meter (kgf * m) to newton meter (N * m)

kilogram-force per square centimeter (kgf/cm^2)
   to pascal (Pa)

kilogram-force per square centimeter (kgf/cm^2)
   to kilopascal (kPa)

kilogram-force per square meter (kgf/m^2) to pascal (Pa)

kilogram-force per square millimeter (kgf/mm^2)
   to pascal (Pa)

kilogram-force per square millimeter (kgf/mm^2)
   to megapascal (Pa)

kilogram-force second squared per meter (kgf * s^2/m)
   to kilogram (kg)

*kilometer per hour (km/h) to meter per second (m/s)

kilopond (kilogram-force) (kp) to newton (N)

*kilowatt hour (kW * h) to joule (J)

*kilowatt hour (kW * h) to megajoule (MJ)

Major/Minor Tab   Unidirectional Conversion Factors
A-Z/I-K

     kip (1 kip = 1 000 lbf) to newton (N)

     kip (1 kip = 1 000 lbf) to kilonewton (kN)

     kip per square inch (ksi) (kip/in^2) to pascal (Pa)

     kip per square inch (ksi) (kip/in^2) to kilopascal (kPa)

     *knot (nautical mile per hour) to meter per second (m/s)

A-Z / L-M

     lambert to candela per square meter (cd/m^2)

     langley (cal th/cm^2) to joule per square meter (J/m^2)

     light year (l.y.) to meter (m)

     *liter (L) to cubic meter (m^3)

     lumen per square foot (lm/ft^2) to lux (lx)

     maxwell (Mx) to weber (Wb)

     mho to siemens (S)

     microinch to meter (m)

     microinch to micrometer (um)

     micron (u) to meter (m)

     micron (u) to micrometer (um)

     mil (0.001 in) to meter (m)

     mil (0.001 in) to millimeter (mm)

     mil (angle) to radian (rad)

     mil (angle) to degree, mile (mi) to meter (m)

     mile (mi) to kilometer (km)

     mile (based on US survey foot) (mi) to meter (m)

     mile (based on US survey foot) (mi) to kilometer (km)

     *mile, nautical to meter (m)

     mile per gallon (US) (mpg) (mi/gal)
      to meter per cubic meter (m/m^3)

     mile per gallon (US) (mpg) (mi/gal)
      to kilometer per liter (km/L)

Major/Minor Tab   Unidirectional Conversion Factors
A-Z / L-M

    mile per hour (mi/h) to meter per second (m/s)

    mile per hour (mi/h) to kilometer per hour (km/h)

    mile per minute (mi/min) to meter per second (m/s)

    mile per second (mi/s) to meter per second (m/s)

    millibar (mbar) to pascal (Pa)

    millibar (mbar) to kilopascal (kPa)

    *millimeter of mercury, conventional (mmHg)
      to pascal (Pa)

    millimeter of water, conventional (mmH2O) to pascal (Pa)

    *minute (angle) (') to radian (rad)

    minute (min) to second (s)

    minute (sidereal) to second (s)

A-Z / N-O

    oersted (Oe) to ampere per meter (A/m)

    *ohm centimeter (ohm * cm) to ohm meter (ohm * m)

    ohm circular-mil per foot to ohm meter (ohm * m)

    ohm circular-mil per foot
      to ohm square millimeter per meter (ohm * mm^2/m)

    ounce (avoirdupois) (oz) to kilogram (kg)

    ounce (avoirdupois) (oz) to gram (g)

    ounce (troy or apothecary) (oz) to kilogram (kg)

    ounce (troy or apothecary) (oz) to gram (g)

    ounce [Canadian and U.K. fluid (Imperial)] (fl oz)
      to cubic meter (m^3)

    ounce [Canadian and U.K. fluid (Imperial)] (fl oz)
      to milliliter (mL)

    ounce (U.S. fluid) (fl oz) to cubic meter (m^3)

    ounce (U.S. fluid) (fl oz) to milliliter (mL)

    ounce (avoirdupois)-force (ozf) to newton (N)

Major/Minor Tab   Unidirectional Conversion Factors
A-Z / N-O

ounce (avoirdupois)-force inch (ozf * in)
  to newton meter (N * m)

ounce (avoirdupois)-force inch (ozf * in)
  to millinewton meter (mN * m)

ounce (avoirdupois) per cubic inch (oz/in^3)
  to kilogram per cubic meter (kg/m^3)

ounce (avoirdupois) per gallon
            [Canadian and U.K. (Imperial)] (oz/gal)
  to kilogram per cubic meter (kg/m^3)

ounce (avoirdupois) per gallon
            [Canadian and U.K. (Imperial)] (oz/gal)
  to gram per liter (g/L)

ounce (avoirdupois) per gallon (U.S.) (oz/gal)
  to kilogram per cubic meter (kg/m^3)

ounce (avoirdupois) per gallon (U.S.) (oz/gal)
  to gram per liter (g/L)

ounce (avoirdupois) per square foot (oz/ft^2)
  to kilogram per square meter (kg/m^2)

ounce (avoirdupois) per square inch (oz/in^2)
  to kilogram per square meter (kg/m^2)

ounce (avoirdupois) per square yard (oz/yd^2)
  to kilogram per square meter (kg/m^2)

A-Z / P-Po

parsec (pc) to meter (m)

peck (U.S.) (pk) to cubic meter (m^3)

peck (U.S.) (pk) to liter (L)

pennyweight (dwt) to kilogram (kg)

pennyweight (dwt) to gram (g)

perm (0 degC) to kilogram per pascal second square meter
            [kg / (Pa * s * m^2)]

Major/Minor Tab **Unidirectional Conversion Factors**
A-Z / P-Po

perm (23 degC) to kilogram per pascal second square meter
[kg / (Pa * s * m^2)]

perm inch (0 degC) to kilogram per pascal second meter
[kg / (Pa * s * m)]

perm inch (23 degC) to kilogram per pascal second meter
[kg / (Pa * s * m)]

phot (ph) to lux (lx)

pica (computer) (1/6 in) to meter (m)

pica (computer) (1/6 in) to millimeter (mm)

pica (printers) to meter (m)

pica (printers) to millimeter (mm)

pint (U.S. dry) (dry pt) to cubic meter (m^3)

pint (U.S. dry) (dry pt) to liter (L)

pint (U.S. liquid) (liq pt) to cubic meter (m^3)

pint (U.S. liquid) (liq pt) to liter (L)

point (computer) (1/72 in) to meter (m)

point (computer) (1/72 in) to millimeter (mm)

point (printers) to meter (m)

point (printers) to millimeter (mm)

poise (P) to pascal second (Pa * s)

pound (avoirdupois) (lb) to kilogram (kg)

pound (troy or apothecary) (lb) to kilogram (kg)

poundal to newton (N)

poundal per square foot to pascal (Pa)

poundal second per square foot to pascal second (Pa * s)

pound foot squared (lb * ft^2)
to kilogram meter squared (kg * m^2)

pound-force (lbf) to newton (N)

pound-force foot (lbf * ft) to newton meter (N * m)

pound-force foot per inch (lbf * ft/in)
to newton meter per meter (N * m/m)

Major/Minor Tab   Unidirectional Conversion Factors
A-Z / Po-Pz

pound-force inch (lbf ° in) to newton meter (N ° m)

pound-force inch per inch (lbf ° in/in)
    to newton meter per meter (N ° m/m)

pound-force per foot (lbf/ft) to newton per meter (N/m)

pound-force per inch (lbf/in) to newton per meter (N/m)

pound-force per pound (lbf/lb) (thrust to mass ratio)
    to newton per kilogram (N/kg)

pound-force per square foot (lbf/ft^2) to pascal (Pa)

pound-force per square inch (psi) (lbf /in^2) to pascal (Pa)

pound-force per square inch (psi) (lbf /in^2)
    to kilopascal (kPa)

pound-force second per square foot (lbf ° s/ft^2)
    to pascal second (Pa ° s)

pound-force second per square inch (lbf ° s/in^2)
    to pascal second (Pa ° s)

pound inch squared (lb ° in^2)
    to kilogram meter squared (kg ° m^2)

pound per cubic foot (lb/ft^3)
    to kilogram per cubic meter (kg/m^3)

pound per cubic inch (lb/in^3)
    to kilogram per cubic meter (kg/m^3)

pound per cubic yard (lb/yd^3)
    to kilogram per cubic meter (kg/m^3)

pound per foot (lb/ft) to kilogram per meter (kg/m)

pound per foot hour [lb/(ft ° h)] to pascal second (Pa ° s)

pound per foot second [lb/(ft ° s)] to pascal second (Pa ° s)

pound per gallon [Canadian and U.K. (Imperial)] (lb/gal)
    to kilogram per cubic meter (kg/m^3)

pound per gallon [Canadian and U.K. (Imperial)] (lb/gal)
    to kilogram per liter (kg/L)

Major/Minor Tab   **Unidirectional Conversion Factors**
A-Z / Po-Pz

pound per gallon (U.S.) (lb/gal)
   to kilogram per cubic meter (kg/m^3)

pound per gallon (U.S.) (lb/gal)
   to kilogram per liter (kg/L)

pound per horsepower hour [lb/(hp * h)]
   to kilogram per joule (kg/J)

pound per hour (lb/h) to kilogram per second (kg/s)

pound per inch (lb /in) to kilogram per meter (kg/m)

pound per minute (lb/min) to kilogram per second (kg/s)

pound per second (lb/s) to kilogram per second (kg/s)

pound per square foot (lb/ft^2)
   to kilogram per square meter (kg/m^2)

pound per square inch (NOT pound-force) (lb/ft^2)
   to kilogram per square meter (kg/m^2)

pound per yard (lb/yd) to kilogram per meter (kg/m)

psi (pound-force per square inch) (lbf /in^2) to pascal (Pa)

psi (pound-force per square inch) (lbf /in^2)
   to kilopascal (kPa)

A-Z / Q-R

quad (10^15 Btu IT) to joule (J)

quart (U.S. dry) (dry qt) to cubic meter (m^3)

quart (U.S. dry) (dry qt) to liter (L)

quart (U.S. liquid) (liq qt) to cubic meter (m^3)

quart (U.S. liquid) (liq qt) to liter (L)

*rad (absorbed dose) (rad) to gray (Gy)

*rem (rem) to sievert (Sv)

revolution (r) to radian (rad)

revolution per minute (rpm) (r/min)
   to radian per second (rad/s)

Major/Minor Tab	Unidirectional Conversion Factors
A-Z / Q-R	
	rhe to reciprocal pascal second (Pa * s)^-1
	rod (based on U.S. survey foot) (rd) to meter (m)
	*roentgen (R) to coulomb per kilogram (C/kg)
	rpm (revolution per minute) (r/min) to radian per second (rad/s)
A-Z / S	
	*second (angle) (") to radian (rad)
	second (sidereal) to second (s)
	shake to second (s)
	shake to nanosecond (ns)
	slug (slug) to kilogram (kg)
	slug per cubic foot (slug/ft^3) to kilogram per cubic meter (kg/m^3)
	slug per foot second (slug/(ft * s) to pascal second (Pa * s)
	square foot (ft^2) to square meter (m^2)
	square foot per hour (ft^2 / h) to square meter per second (m^2/s)
	square foot per second (ft^2/s) to square meter per second (m^2/s)
	square inch (in^2) to square meter (m^2)
	square inch (in^2) to square centimeter (cm^2)
	square mile (mi^2) to square meter (m^2)
	square mile (mi^2) to square kilometer (km^2)
	square mile (based on U.S. survey foot) (mi^2) to square meter (m^2)
	square mile (based on U.S. survey foot) (mi^2) to square kilometer (km^2)
	square yard (yd^2) to square meter (m^2)

Major/Minor Tab   Unidirectional Conversion Factors
A-Z / S

statampere to ampere (A)

statcoulomb to coulomb (C)

statfarad to farad (F)

stathenry to henry (H)

statmho to siemens (S)

statohm to ohm (ohm)

statvolt to volt (V)

stere (st) to cubic meter (m^3)

stilb (sb) to candela per square meter (cd / m^2)

stokes (St) to meter squared per second (m^2 / s)

A-Z / T-Z

tablespoon to cubic meter (m^3)

tablespoon to milliliter (mL)

teaspoon to cubic meter (m^3)

teaspoon to milliliter (mL)

tex to kilogram per meter (kg / m)

therm (EC) to joule (J)

therm (U.S.) to joule (J)

ton, assay (AT) to kilogram (kg)

ton, assay (AT) to gram (g)

ton-force (2000 lbf) to newton (N)

ton-force (2000 lbf) to kilonewton (kN)

ton, long (2240 lb) to kilogram (kg)

ton, long, per cubic yard
     to kilogram per cubic meter (kg/m^3)

*ton, metric to kilogram (kg)

tonne (called metric ton in U.S.) (t) to kilogram (kg)

ton of refrigeration (12 000 Btu IT/h) to watt (W)

ton of TNT (energy equivalent) to joule (J)

ton, register to cubic meter (m^3)

ton, short (2000 lb) to kilogram (kg)

Major/Minor Tab **Unidirectional Conversion Factors**
A-Z / T-Z

ton, short, per cubic yard
to kilogram per cubic meter (kg/m^3)

ton, short, per hour to kilogram per second (kg/s)

torr (Torr) to pascal (Pa)

unit pole to weber (Wb)

*watt hour (W ° h) to joule (J)

*watt per square centimeter (W/cm^2)
to watt per square meter (W/m^2)

watt per square inch (W /in^2)
to watt per square meter (W/m^2)

*watt second (W ° s) to joule (J)

yard (yd) to meter (m)

*year (365 days) to second (s)

year (sidereal) to second (s)

year (tropical) to second (s)

CALCULATION

d. Enter the "from" value into the edit field and click the Enter button.

The number of significant digits in the "From" value controls the default number of significant digits that will be displayed when the "To" value has been evaluated. Please see the section on Significant Digits below for more information and a means of overriding the default value if desired.

e. The Convert button will activate when the app has verified that the "From" value is in a valid format and has been checked for significant digits. Click the Convert button to calculate the "To" value and to display the conversion factor and its exact/derived status.

Note that the Temperature conversions do not have multiplicative conversion factors but are calculated using formulas, so the app confirms that the conversion has been calculated rather than displaying a conversion factor.

CONVERSION EXAMPLES

CONVERSION EXAMPLE 1 – MILES TO FEET

Suppose you have a distance measurement in miles that was measured to four significant digits and you need to convert it to feet.

1. Click only one time on the app icon in the Apps toolbar. The control panel will open on the computer screen (see Fig C.1).

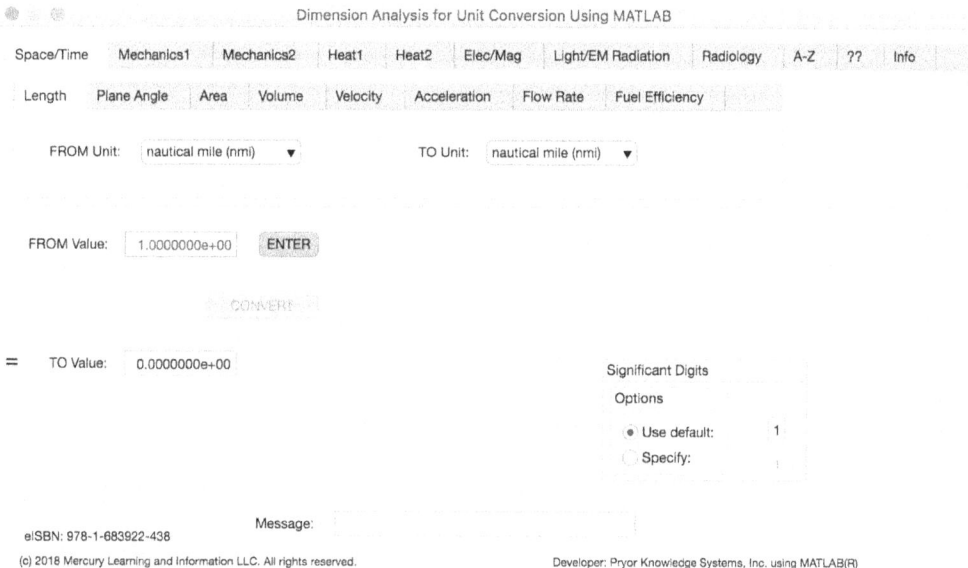

Dimension Analysis for Unit Conversion Using MATLAB										
Space/Time	Mechanics1	Mechanics2	Heat1	Heat2	Elec/Mag	Light/EM Radiation	Radiology	A-Z	??	Info
Length	Plane Angle	Area	Volume	Velocity	Acceleration	Flow Rate	Fuel Efficiency			

FROM Unit: nautical mile (nmi) ▾   TO Unit: nautical mile (nmi) ▾

FROM Value: 1.0000000e+00 ENTER

CONVERT

= TO Value: 0.0000000e+00

Significant Digits

Options

● Use default: 1

○ Specify:

Message:

Developer: Pryor Knowledge Systems, Inc. using MATLAB(R)

FIGURE C.1 Opening App Screen.

2. Click on the major category tab, Space/Time, and then click the Length tab.

3. Select the value and units to be converted:

    a. FROM Unit is mile (mi)

    b. TO Unit is foot (ft)

    c. Enter the From Value as 1.000

    d. Click the Enter Button. The default number of significant digits is 1.

    e. Set the Significant Digits option to Specify and the number to 4.

    f. Click the Enter button and then the Convert button. The Conversion Factor displays as 5.2800000E+03, and the "Exact" lamp is lit to indicate that this factor has been defined precisely,

    g. The TO Value displays as 5.2800000E+03(See Fig C.2).

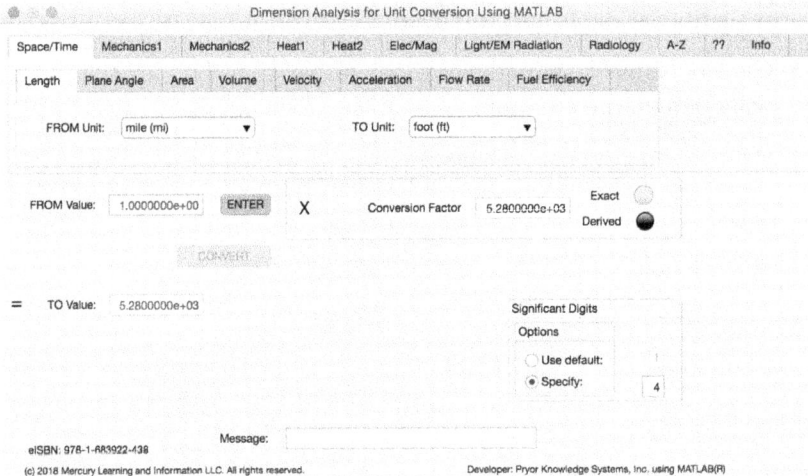

FIGURE C.2 Length screen after conversion.

4. To continue with more length conversions, change the units and enter the new From value, click the Enter button, and proceed as before. To reset the panel selections, click another tab, e.g., the Plane Angle tab, and then click the Length tab. *Note that the pull-down unit selectors reset to the first item in the list and the From, To, and Significant Digits return to their default values.*

CONVERSION EXAMPLE 2 – STANDARD ACCELERATION OF GRAVITY TO METERS PER SECOND SQUARED

In this case, you have g, the standard acceleration of gravity on planet Earth, and you need to express this value in SI units, m/s², with six significant digits.

1. Click on the major category tab, Space/Time, and then on the Acceleration tab (See Fig C.1).

2. Select the value and units to be converted, e.g.,

    a. FROM Unit is standard acceleration of gravity (g)

    b. TO Unit is meters per second squared (m/s^2)

    c. Enter the From Value as 1.000

    d. Click the Enter Button. The default number of significant digits is set to 1.

    e. Set the Significant Digits option to Specify and the number to 6.

    f. Click the Enter button and then the Convert button. The Conversion Factor displays as 9.8066500E+00, and the "Exact" lamp is lit to indicate that this conversion factor has been defined precisely.

    g. The TO Value displays as 9.8066500E+00 (See Fig C.3).

FIGURE C.3 Acceleration screen after conversion.

CONVERSION EXAMPLE 3 – PINT TO CUBIC METER

In this case, you have a pint (liq pt) value (2.0), and you need to express this value in the SI unit, cubic meters (m^3), with seven significant digits.

1. Click on the major category tab, A – Z, and then on the P-Po tab (see Fig. C.4).

2. Select the value and units to be converted, e.g.,

a. the UNITS FROM-->TO selection is pint (U.S. liquid) (liq pt) --> cubic meter (m^3)

b. Enter the FROM value as 2

c. Click the Enter button. The default number of significant digits is set to 2.

d. Set the Significant Digits option to Specify and the number to 7.

e. Click the Enter button and then the Convert button. The Conversion Factor displays as 4.731765e-04, and the Derived lamp is lit to indicate that this conversion factor has not been defined precisely.

f. The TO Value displays as 9.46353000e-04. (See Fig. C.4.)

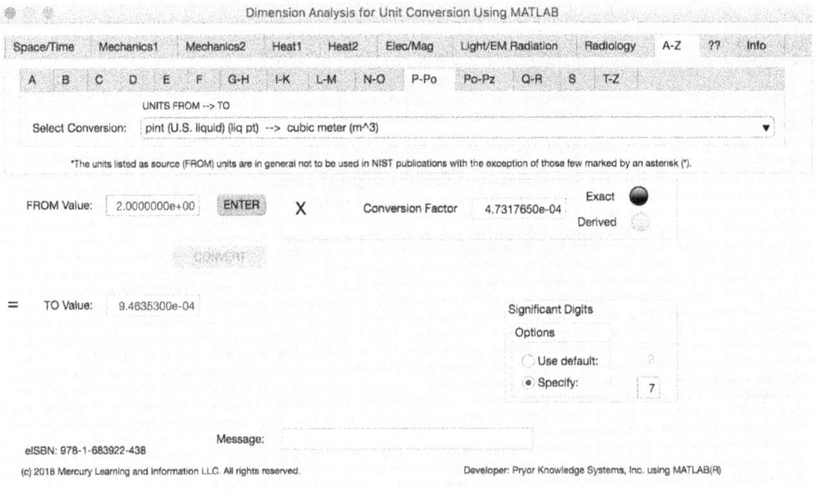

FIGURE C.4 Screen A-Z/P-Po after conversion.

SUPPORT TABS AND HELP

??(HELP) TAB

The help tab contains the sections of this user guide for ready reference.

INFO (ABOUT) TAB

The About tab sections contain the copyright information, the license, liability, and warranty terms, and information about the developer.

SIGNIFICANT DIGITS

The number of significant digits determines rounding of the result of the conversion calculation in order to retain the precision of the original measurement.

DEFAULT NUMBER OF SIGNIFICANT DIGITS

The number of significant digits in the "From" value controls the default number of significant digits that will be displayed when the "To" value has been calculated. When the "From" value is entered, the value is examined to determine how many of the digits entered are necessary to retain the value when extraneous zeros are removed. Numbers with trailing zeros to the left of the decimal point assume that the zeros indicate an exact measurement. For example, the app treats an entry of 1000 as four significant digits while an entry of 999 would have three significant digits. The default number of significant digits is limited to a maximum of five (5) digits because the number of significant digits in the conversion factors varies from five to eight digits.

If the resulting "To" value has a lower first digit than the "From" value, the app adds an additional significant digit to the "To" value to account for a possible increase in magnitude resulting from the conversion calculation. For example, to convert 66 miles to kilometers, the calculation 66 x 1.609347 will equal 106.2169 kilometers. The first significant digit of the result (1) is lower than the first digit of the inch-pound value (6). Therefore, the converted value should have one more significant digit than the original value and the result is 106 km.

OVERRIDING THE NUMBER OF SIGNIFICANT DIGITS

If you know that the accuracy of the "From" value would be better represented by a different number of digits, you can override the default number of significant digits. Select the "Specify" option and enter the number of significant digits that should apply to the conversion calculation up to a maximum of seven (7). For example, if an entry of 1000 is the result of an estimate to the nearest thousand rather than a measurement, an override value of 1 significant digit would be more accurate.

REFERENCES

1. "The International System of Units (SI) – Conversion Factors for General Use," National Institute of Standards and Technology Special Publications 1038, Natl. Inst. Stand. Technol. Spec. Pub. 1038, 24 pages (May 2006).

2. "Guide for the Use of the International System of Units (SI)," National Institute of Standards and Technology Special Publication 811, 2008 Ed., 85 pages (March 2008, 2nd printing November 2008).

INDEX

www.ingramcontent.com/pod-product-compliance
Lightning Source LLC
Chambersburg PA
CBHW081531220326
41598CB00036B/6396